Discovering Discrete Dynamical Systems

Library of Congress Catalog Card Number 2017939374

Print edition ISBN 978-0-88385-793-9

Electronic edition ISBN 978-1-61444-124-3

Printed in the United States of America

Current Printing (last digit):
10 9 8 7 6 5 4 3 2 1

Discovering Discrete Dynamical Systems

Aimee Johnson
Swarthmore College

Kathleen Madden
California State University, Bakersfield

Ayşe Şahin
Wright State University

Published and Distributed by
The Mathematical Association of America

CLASSROOM RESOURCE MATERIALS

Classroom Resource Materials is intended to provide supplementary classroom material for students—laboratory exercises, projects, historical information, textbooks with unusual approaches for presenting mathematical ideas, career information, etc.

101 Careers in Mathematics, 3rd edition edited by Andrew Sterrett

Archimedes: What Did He Do Besides Cry Eureka?, Sherman Stein

Arithmetic Wonderland, Andrew C. F. Liu

Calculus: An Active Approach with Projects, Stephen Hilbert, Diane Driscoll Schwartz, Stan Seltzer, John Maceli, and Eric Robinson

Calculus Mysteries and Thrillers, R. Grant Woods

Conjecture and Proof, Miklós Laczkovich

Counterexamples in Calculus, Sergiy Klymchuk

Creative Mathematics, H. S. Wall

Discovering Discrete Dynamical Systems, Aimee Johnson, Kathleen Madden, and Ayşe Şahin

Environmental Mathematics in the Classroom, edited by B. A. Fusaro and P. C. Kenschaft

Excursions in Classical Analysis: Pathways to Advanced Problem Solving and Undergraduate Research, by Hongwei Chen

Explorations in Complex Analysis, Michael A. Brilleslyper, Michael J. Dorff, Jane M. McDougall, James S. Rolf, Lisbeth E. Schaubroeck, Richard L. Stankewitz, and Kenneth Stephenson

Exploratory Examples for Real Analysis, Joanne E. Snow and Kirk E. Weller

Exploring Advanced Euclidean Geometry with GeoGebra, Gerard A. Venema

Game Theory Through Examples, Erich Prisner

Geometry From Africa: Mathematical and Educational Explorations, Paulus Gerdes

The Heart of Calculus: Explorations and Applications, Philip Anselone and John Lee

Historical Modules for the Teaching and Learning of Mathematics (CD), edited by Victor Katz and Karen Dee Michalowicz

Identification Numbers and Check Digit Schemes, Joseph Kirtland

Interdisciplinary Lively Application Projects, edited by Chris Arney

Introduction to the Mathematics of Computer Graphics, Nathan Carter

Inverse Problems: Activities for Undergraduates, Charles W. Groetsch

Keeping it R.E.A.L.: Research Experiences for All Learners, Carla D. Martin and Anthony Tongen

Laboratory Experiences in Group Theory, Ellen Maycock Parker

Learn from the Masters, Frank Swetz, John Fauvel, Otto Bekken, Bengt Johansson, and Victor Katz

Math Made Visual: Creating Images for Understanding Mathematics, Claudi Alsina and Roger B. Nelsen

Mathematics Galore!: The First Five Years of the St. Marks Institute of Mathematics, James Tanton

MAA Service Center
P.O. Box 91112
Washington, DC 20090-1112
1-800-331-1MAA FAX: 1-301-206-9789

Ackowledgements

The authors would like to thank their students at DePaul University, Drew University, and Swarthmore College who helped pilot these modules and contributed greatly to their improvement. Special thanks to Harriet Pollatsek, Andrew Dykstra, Ed Kroc, Michelle LeMasurier, and Aidan McCaffrey for their feedback, and to Steve Maurer for his help in bringing this textbook forward to publication.

To the Instructor

Discovering Discrete Dynamical Systems is designed for use in a student-led, discovery-based course for advanced mathematics majors. When used as intended, the structure of this text will enable students to develop persistence and skill in exploration, conjecture, and generalization; to apply principles of real analysis to the study of dynamical systems; to read mathematics independently; and to communicate mathematical ideas with clarity.

The organization of the text into modules is crucial to its success in meeting our objectives. Each module will take three or four class periods to cover, depending on the length of time of a class meeting. The first three modules lay the foundation for all the material that follows. After these modules, the instructor has a great deal of freedom in choosing the topics for the remainder of the course. The table below gives the prerequisites for each module to help with course planning. In addition, Modules 9 and 10 assume familiarity with basic linear algebra and complex numbers respectively. Familiarity with the notion of cardinality is helpful, but not required, for Module 4.

Module 4:	Modules 1–3
Module 5:	Modules 1–3
Module 6:	Modules 1–3
Module 7:	Modules 1–3
Module 8:	Modules 1–3, project requires Module 5
Module 9:	Modules 1–3, project requires Module 7
Module 10:	Modules 1–4
Module 11:	Modules 1–3, 8
Module 12:	Modules 1–4, 7, 8
Module 13:	Modules 1–3
Module 14:	Modules 1–4, 8, 13

Table 1. Module Prerequisites.

Each module starts with an exploration in which the students are asked an open-ended question. The goal of the exploration is to allow the students to make discoveries that lead them to formulate the questions addressed in the exposition of the module. Each exploration can be covered in a class period, with the first half of the period spent with the students exploring in groups. During the remainder of the period, the groups can report their findings and the class can discuss as a whole any questions, ideas, and conjectures inspired by the student discoveries. The text is careful to never follow an exploration with an exposition that overtly answers the questions raised in the exploration.

The module exposition follows the exploration. This exposition is brief, and any student who has the prerequisites for an analysis course can read it independently. When the text is used as intended, students successfully read the exposition, not only because in the absence of lectures this is the only source of information that they have, but also because the explorations have previously engaged them in the material. The module exposition is followed by exercises designed to reinforce and build upon the reading. The exposition and exercises can be assigned as homework with the students working independently or in groups. The next class period can then be spent with the students asking questions about the exposition and presenting and discussing homework solutions.

Each module concludes with a project that is designed to bring the ideas from the module to bear on a more challenging or in-depth problem. There are many ways instructors can use the projects: for example, students can work in groups on all of the projects, with one group assigned to lead the project discussion; the projects can be divided up between student groups with each group presenting their own project; or some combination of these approaches can be used.

The course can end with students investigating a subject of their own and presenting a brief description to the class. Students can use their own interests to decide on the topics but may also be encouraged to search for relevant papers in *The College Mathematics Journal* or *The American Mathematical Monthly*, for instance.

The explorations and projects will require the use of technology, particularly for the iteration of functions. Past instructors have used various approaches: some have provided the basic coding for their students, using Mathematica, Maple, or Excel. Others have had the students do their own programming, while others have taken advantage of the many web-based applications available on-line.

When using this text for a discovery-based course, the most important thing to do is also the hardest: instructors should step back and let students grapple with their ideas and questions with very little instructor involvement. Students might not learn as many facts on their own as they would with instruction, but they will experience the challenge and thrill of mathematical discovery. They will overcome uncertainty and frustration and grow in confidence and self-sufficiency. To witness that growth is a teacher's best reward!

To the Student

Fermat's Last Theorem, a bedeviling open problem in mathematics that remained unsolved for over 300 years, was finally proved in 1995 by Princeton mathematician Andrew Wiles. Professor Wiles describes the work of doing research in mathematics as analogous to exploring a dark mansion [Wi]:

> You enter the first room of the mansion and it's completely dark. You stumble around bumping into the furniture but gradually you learn where each piece of furniture is. Finally, after six months or so, you find the light switch, you turn it on, and suddenly it's all illuminated. You can see exactly where you were. Then you move into the next room and spend another six months in the dark. So each of these breakthroughs, while sometimes they're momentary, sometimes over a period of a day or two, they are the culmination of, and couldn't exist without, the many months of stumbling around in the dark that precede them.

"Stumbling around in the dark" is probably not the way you have experienced mathematics. If you are in this course, you are good at mathematics and you have studied quite a lot of it. Your previous mathematics courses have exposed you to various fields of mathematics (e.g., calculus, linear algebra, etc.) that are fully established; you have struggled to master their principles and techniques; and because you study hard, you have been able to use them to arrive at the correct answers to specific, focused questions.

Your previous experience is important and foundational, however it is not the experience of a mathematician doing mathematical research. When doing research, the answers are unknown, and in fact, the right questions to answer may not even be clear. Although mathematicians bring the skills and techniques they have learned previously to every new problem, it also might not be clear which skills and techniques will be helpful, or if new skills and new techniques must be developed.

This book is designed to allow you to experience mathematical discovery in the way that mathematicians do, in the way described by Andrew Wiles.

Each module in this book begins with an exploration. The explorations are open-ended questions posed with little or no guidance on how to proceed. You may feel frustrated when you do not know where to begin or when you are unsure exactly what is required of you; this is the experience of stumbling around a darkened room. In the course of struggling with the mathematical ideas in the modules, you will gain important skills — to come up with appropriate questions and conjectures, to test ideas with examples, to distinguish between a proof and intuition, and to work as part of a team — so that over time you will become more comfortable with openness and uncertainty.

Following each exploration is a brief exposition and a set of exercises. They are designed to build on the intuition and understanding you have developed in the exploration. The answers to many of the questions you grappled with in the exploration will become clear after you read the exposition and work your way through the exercises.

The exposition and exercises will also increase your skills as an independent reader of mathematics. As you know, reading mathematics is not like reading in other disciplines; you must read slowly, trying examples and testing your understanding of theorems and definitions. Some of the questions found in the exercises are designed to strengthen your mathematical reading skills.

Each module concludes with a project. Like the explorations, the projects involve open-ended questions and little guidance. As in the exploration, you may feel that you are again stumbling around in a dark room, but you will find that you bring a greater depth of understanding and experience to the project as a result of your earlier work in the module.

If you feel unsure and unconfident when learning mathematics in this new, independent, less structured way, you should know that is a normal response. Don't give up! You will be surprised at the skills you will learn, skills that will be applicable to any problem-solving exercise (mathematical or not) in your future. With hard work, you will also find the light switch in some of the dark rooms you have been exploring in the mansion of discrete dynamical systems, and as any mathematician will tell you, there is no more satisfying feeling. So, open yourself up to something new, empowering, and exciting, and enjoy the journey!

Contents

Module 1

Fixed Points of Dynamical Systems

Exploration

Consider the function

$$g(x) = 2x(1 - x)$$

where x is a real number. Pick an initial value for x, denoted x_0, and iterate the function. Namely, find $g(x_0)$, $g^2(x_0) = g(g(x_0))$, $g^3(x_0) = g(g(g(x_0)))$, and so on. Do this for a variety of initial values and try to find all possible long term behaviors of these initial values under iteration.

The function g belongs to the **logistic family** of functions defined by

$$g_c(x) = cx(1 - x).$$

Try changing c to various other values and discuss how this changes the behavior of initial values of x under iterates of the function g_c.

Exposition

One of the goals of science is to predict the future based on current conditions. If today the planet Mars is visible at 8 P.M. over the northwestern horizon, will it be visible at 8 P.M. over the same horizon a month from now? If a company's stock is $23 a share today, should we sell it? If 10% of a petri dish is currently covered by bacteria, when will it be completely covered?

Scientists often try to answer these sorts of questions by finding a mathematical model that describes the system. For example, suppose the fraction of a petri dish covered by bacteria seems to double every hour. If we let x_0 represent the fraction of the dish covered at some moment in time, then after one hour the fraction covered is $2x_0$, after two hours the fraction covered is $2(2x_0)$, and so on. We can try to model this situation by using the function $h(x) = 2x$ where x is a real number in $[0, 1]$. If the initial fraction of the dish covered is x_0, then we can denote the amount covered after one hour by $h(x_0)$, the amount covered after two hours by $h^2(x_0) = h(h(x_0))$, and so on.

This model clearly has problems. In particular, it does not take into account the fact that as the petri dish fills up, crowding begins to affect the growth of bacteria. We can improve our model by using the function $g(x) = 2x(1-x)$. When x is small, the $(1-x)$ term is close to one and the amount of bacteria nearly doubles every hour. But if x is closer to one, the $(1-x)$ term is less than one half and the population will decrease.

This is an example of a **dynamical system**. A dynamical system consists of a pair (X, f) where X is the set of all possible states of the system, and $f : X \to X$ is a function which models how the system will evolve over time given an initial starting point in X. In the example, the points in $X = [0, 1]$ are all the possible fractions of the petri dish covered by bacteria and the function is $g(x) = 2x(1-x)$. X is called the **phase space** of the dynamical system and the fraction of the dish covered by bacteria at the time of measurement is called an **initial condition** or **initial value**. In this book, unless otherwise noted, the phase space will be a subset of $\mathbb{R}$.

The main goal in the study of a dynamical system (X, f) is to describe the long term, or asymptotic, behavior of the points in X under the time evolution rule given by f. In our example, if we want to understand the long term behavior of the bacteria for different initial amounts, we need to understand the effect of repeatedly applying, or iterating, the function g to different choices of initial conditions x_0; that is, we need to study the set

$$\{x_0, g(x_0), g^2(x_0), g^3(x_0), \cdots, g^n(x_0), \cdots\}$$

for different choices of x_0. For a point $x_0 \in X$, this set of iterates is called the **orbit of x_0**.

Some initial conditions have a simple orbit consisting of a single point. These initial conditions play an important role in the study of a dynamical system.

Definition 1.1. *A point p in the domain of a function f is called a* **fixed point** *of f if $f(p) = p$. It is called* **eventually fixed** *if one of its iterates is a fixed point; namely, if there is some n so that $f(f^n(p)) = f^n(p)$.*

As seen in the exploration, there are different types of fixed points. Numerical experiments will have shown that for certain fixed points p, if we iterate nearby initial conditions x_0, then as we take more and more iterates, they get closer and closer to p. In fact, the iterates form a convergent sequence with limit p:

Definition 1.2. *A sequence of numbers $\{x_n\}$ is said to* **converge** *to a point x^* if for all $\epsilon > 0$ there is a positive integer $N \in \mathbb{N}$, such that for all integers $n \geq N$,*

$$\left|x_n - x^*\right| < \epsilon.$$

The point x^* is called the **limit** *of the sequence and we write $\lim_{n\to\infty} x_n = x^*$.*

The type of fixed point described above can now be defined as follows.

Definition 1.3. *Let $p \in X$ be a fixed point of function f. We say p is an* **attracting fixed point** *if there exists $\epsilon > 0$ such that every $x \in X$ with $|x - p| < \epsilon$ satisfies $\lim_{n\to\infty} f^n(x) = p$.*

For some other fixed points, the opposite seems to happen: nearby initial conditions move farther away from p under iteration. These are called repelling fixed points.

Definition 1.4. *Let $p \in X$ be a fixed point of a function f. We say p is a* **repelling fixed point** *if there exists $\epsilon > 0$ such that for every $x \in X$ with $|x - p| < \epsilon$ and $x \neq p$, there exists $k \in \mathbb{N}$ such that $\left|f^k(x) - p\right| \geq \epsilon$.*

We say an initial condition p is a (repelling, attracting) fixed point of a dynamical system (X, f) if $p \in X$ is a (repelling, attracting) fixed point of f.

Let p be an attracting fixed point of a dynamical system (X, f). The set of all initial conditions x_0 with the property that the sequence $\{f^n(x_0)\}$ converges to p is called the **basin of attraction** of p. The definition of an attracting fixed point guarantees that there is an interval containing p in the basin of attraction. However, as we will see in the next module, the basin of attraction can consist of more than a single interval.

Fixed points of (X, f) are solutions to the equation $f(x) = x$. Geometrically, they are the intersections of the graph of f with the graph of the line $y = x$. Figure 1.1 illustrates the two fixed points of (X, g), the dynamical system described previously.

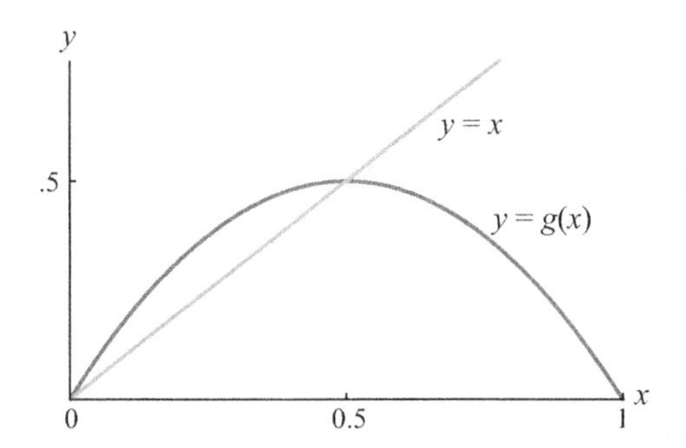

Figure 1.1. The graph of $y = g(x)$ and of $y = x$.

Figure 1.1 can also be used to illustrate iterations of the function g. Because points on the line $y = x$ have coordinates (x, x), we can represent the initial condition $x_0 = 1/10$ by the ordered pair $(1/10, 1/10)$ on the line $y = x$. If we draw a vertical line segment from this point to the graph of g we will be at the point $(1/10, g(1/10))$. Drawing a horizontal line segment from the point $(1/10, g(1/10))$ back to the the line $y = x$, we arrive at the point $(g(1/10), g(1/10))$ that represents the second point in the orbit of $x_0 = 1/10$. This is illustrated in Figure 1.2.

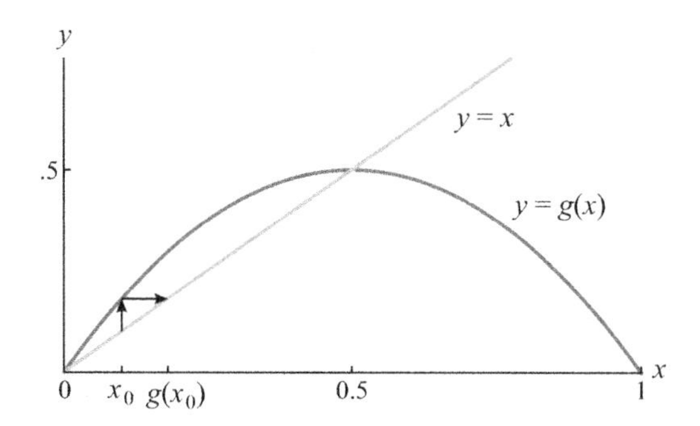

Figure 1.2. An illustration of x_0 and $g(x_0)$ where $x_0 = 1/10$.

Given an initial condition x_0, this procedure gives us a way to obtain a geometric representation of the orbit of x_0 by using each point $(g^n(x_0), g^n(x_0))$ on the line $y = x$ to represent $g^n(x_0)$. Figure 1.3 illustrates a geometric representation of $g(1/10)$, $g^2(1/10)$, $g^3(1/10)$, and $g^4(1/10)$, the first four iterates of $1/10$. Diagrams such as these are referred to as **web diagrams**.

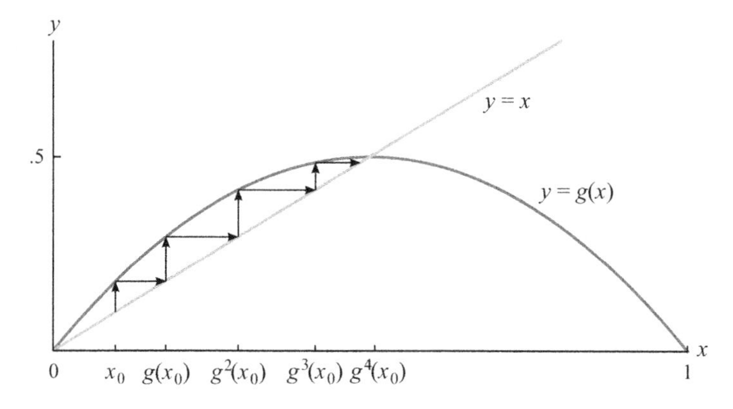

Figure 1.3. An illustration of $\{x_0, g(x_0), g^2(x_0), g^3(x_0), g^4(x_0)\}$ where $x_0 = 1/10$.

Web diagrams and numerical experiments such as those from the exploration are useful tools to give us insight into the long term behavior of initial conditions for a dynamical system. For example, Figure 1.3, a few more web diagrams, and the numerical iterations from the exploration lead us to make two conjectures about $x_0 \in (0, 1/2)$. First, the sequence $\{g^n(x_0)\}$ is an increasing sequence not exceeding $1/2$, and second, the sequence converges to the fixed point $1/2$.

The first part of our conjecture is easily checked. Clearly, for $x_0 \in (0, 1/2)$, $g'(x_0) = 2 - 4x_0$ is positive and so g is increasing on $(0, 1/2)$. Furthermore, $g(1/2) = 1/2$, the maximum value of g. Thus iterates of initial conditions in $(0, 1/2)$ increase but must stay in the interval $(0, 1/2)$. A sequence that stays in a bounded interval like this one is said to be **bounded**.

It takes a little more work to verify the second part of our conjecture. First, we must establish that the orbit of an initial condition $x_0 \in (0, 1/2)$ in fact converges. This follows from Theorem 1.5, the proof of which can be found in many calculus textbooks.

Theorem 1.5. (Monotone convergence theorem) *A bounded, monotonic (increasing or decreasing) sequence must converge.*

We have thus established that for $x_0 \in (0, 1/2)$ there exists $p \leq 1/2$ for which $\lim_{n\to\infty} g^n(x_0) = p$. The question now is whether the rest of our conjecture is true. Namely, is there an initial condition in $(0, 1/2)$ whose orbit converges to something less than $1/2$? We will put this question to rest by first establishing that the limit of this sequence must be a fixed point.

Lemma 1.6. *Let f be a continuous function and x_0 an initial condition. Suppose that $\{f^n(x_0)\}$, the iterates of x_0 by f, forms a convergent sequence. Then $\lim_{n\to\infty} f^n(x_0)$ is a fixed point.*

Proof. Let $p = \lim_{n\to\infty} f^n(x_0)$. Notice that

$$f(p) = f\left(\lim_{n\to\infty} f^n(x_0)\right).$$

By the continuity of f,

$$f\left(\lim_{n\to\infty} f^n(x_0)\right) = \lim_{n\to\infty} f\left(f^n(x_0)\right) = \lim_{n\to\infty} f^{n+1}(x_0).$$

But

$$\lim_{n\to\infty} f^{n+1}(x_0) = \lim_{n\to\infty} f^n(x_0) = p$$

so $f(p) = p$ and p is a fixed point. □

The function g in our example certainly satisfies the hypothesis of Lemma 1.6, and thus all initial conditions $x_0 \in (0, 1/2)$ must converge to a fixed point p under iteration. The only two fixed points for the dynamical system (X, g) are 0 and $1/2$. Since $x_0 > 0$ and the sequence $\{g^n(x_0)\}$ is an increasing sequence, we must have that $p = 1/2$.

Web diagrams and numerical experiments from the exploration will also suggest that if x_0 is an initial condition in $(1/2, 1)$, then it will also approach $1/2$ under iteration. This is now easily verified. Since g maps $(1/2, 1)$ onto $(0, 1/2)$, we see that if $x_0 \in (1/2, 1)$ then $g(x_0) \in (0, 1/2)$. We can then use the previous argument to show that iterates of x_0 converge to $1/2$ as well.

In this module, a dynamical system was used to model the growth of bacteria in a petri dish. Dynamical systems arise in many contexts, from modeling the beating human heart to predicting future stock prices. The project will explore an example of a dynamical system arising in an unusual context.

Exercises

1. Create web diagrams for some of the functions g_c and some of the initial conditions you used in the exploration.

2. The following exercises refer to the functions g and g_c as given in the exploration, with their domains restricted to $[0, 1]$.

(a) $([0, 1], g)$ models the growth of bacteria in a petri dish. Interpret your findings from the exploration and Exercise 1 in this context.

(b) For what values of c will the system $([0, 1], g_c)$ be a reasonable model for the growth of bacteria in a petri dish?

(c) For the values of c found in (b), interpret your findings from the exploration and Exercise 1 in the context of bacteria growth in a petri dish.

3. Consider the dynamical system $(\mathbb{R}, g)$ from the exploration.

(a) Find a value for $\epsilon > 0$ satisfying Definition 1.3 for the fixed point $p = 1/2$.

(b) Find a value for $\epsilon > 0$ satisfying Definition 1.4 for the fixed point $p = 0$.

(c) Using the value of ϵ found in (b), pick a positive and negative initial condition from $(-\epsilon, \epsilon)$ and compute the value of k for each.

4. For the following dynamical systems, show that the fixed point $p = 0$ is neither attracting nor repelling by showing that there is no $\epsilon > 0$ satisfying Definition 1.3 or Definition 1.4.

(a) $(\mathbb{R}, h)$ where $h(x) = -x$.

(b) $(\mathbb{R}, g_1)$.

5. An initial condition x_0 is called a **periodic point with period n** for the dynamical system (X, f) if it is a fixed point for (X, f^n). Using technology as needed, find the periodic points of period four for $(\mathbb{R}, g_{3.5})$ and write down their orbits.

6. Give an example of a convergent sequence of numbers $\{x_n\}$ with the property that $x_n < 1$ for all n, but for which $\lim_{n\to\infty} x_n = 1$. Prove your claim. Remember to make use of Definition 1.2.

7. Give an example of an increasing convergent sequence of numbers $\{x_n\}$ with the property that $x_n < 1$ for all n, and for which $\lim_{n\to\infty} x_n < 1$. Prove your claim.

8. Give an example of a bounded sequence $\{x_n\}$ of real numbers that does not converge. Prove your claim.

9. Definition 1.2 says what it means for a sequence $\{x_n\}$ to converge to a value x^*. A sequence **converges to** ∞, written $\lim_{n\to\infty} x_n = \infty$, if for all $K \in \mathbb{N}$ there is an $N \in \mathbb{N}$ such that for all $n \geq N$, $x_n > K$.

(a) Write down the analagous definition for $\lim_{n\to\infty} x_n = -\infty$.

(b) Consider the dynamical system $(\mathbb{R}, g)$ from the exploration.

i. For $x_0 \in (-\infty, 0)$, prove that $\lim_{n\to\infty} g^n(x_0) = -\infty$.

ii. For $x_0 \in (1, \infty)$, prove $\lim_{n\to\infty} g^n(x_0) = -\infty$.

10. Let $h(x) = x^2 + x$ and consider $(\mathbb{R}, h)$.

(a) Prove that for an initial condition x_0 with $-1 < x_0 < 0$, $\lim_{n\to\infty} h^n(x_0) = 0$.

(b) Prove that for an initial condition $x_0 > 0$, $\lim_{n\to\infty} h^n(x_0) = \infty$.

(c) Prove that for an initial condition $x_0 < -1$, $\lim_{n\to\infty} h^n(x_0) = \infty$

(d) What do (7a) and (7b) tell us about the fixed point $p = 0$?

These results will be used in Module 7.

Project

At a Las Vegas extravaganza, the Mathematical Magician displays her numerical powers with the following card trick. She begins by laying her seventeen card deck in a line as shown in Figure 1.4. While the Mathematical Magician leaves the room, a Willing Volunteer chooses a card and shows it to the audience. For example, suppose the Willing Volunteer chooses card number 7. When the Mathematical Magician returns to the room, she collects the cards and arranges them in three columns as in Figure 1.5. The Willing Volunteer indicates that his card is in the first column. The Mathematical Magician then stacks the columns of cards, beginning with one of the longer columns, then with the column indicated by the Willing Volunteer, followed by the remaining column. This is illustrated in Figure 1.6.

Notice that the card chosen by the Willing Volunteer, originally in the seventh position, is now in the ninth position.

The audience is not yet impressed, so the Mathematical Magician continues. She divides the cards into three columns as before, the Willing Volunteer indicates the column containing his card, and the Mathematical Magician again stacks the columns of cards as described above. After repeating this process a third time, the Mathematical Magician announces:

"Your card is seven!"

1
2
3
4
5
6
7
8
9
10
11
12
13
14
15
16
17

Figure 1.4. The first step for the card trick.

16	17	
13	14	15
10	11	12
7	8	9
4	5	6
1	2	3

Figure 1.5. The second step for the card trick.

15
12
9
6
3
16
13
10
7
4
1
17
14
11
8
5
2

Figure 1.6. The third step for the card trick.

The crowd gasps and bursts into spontaneous applause. "How did you do it?"

"It's easy," said the Mathematical Magician, "if you understand the power of attracting fixed points."

How does the Mathematical Magician do it? And can she do it no matter what card the Willing Volunteer picks? To answer this, find a phase space and a time evolution rule modeling the card trick. Are there any attracting fixed points for this process? If so, what are their basins of attraction? When determining the time evolution rule, remember that $f(i)$ will depend on how the cards are arranged by the Mathematical Magician. This arrangment depends on the card i chosen by the Willing Volunteer. In our example, $f(7) = 9$.

Is there a similar magic trick for a deck consisting of nineteen cards? Eighteen cards? Can you determine for what deck sizes the Mathematical Magician can perform her trick?

This project is based on [EH] where many other variations and interesting facts can be found.

Module 2

Classifying Fixed Points

Exploration

Create graphs of $g_c(x) = cx(1-x)$ for a variety of values of c and use them to find the associated fixed points. Can you characterize the number and nature of the fixed points (attracting, repelling, or neither) for different values of c? Is there a relationship between the geometry of the graph and whether the fixed points are attracting, repelling, or neither?

Exposition

When a dynamical system models a physical situation, its fixed points carry important information. First and foremost, they represent equilibrium states of the physical system. In the dynamical system $([0, 1], g_2)$ from the previous module that modeled the growth of bacteria in a petri dish, the point $p = 0$ is a fixed point. This reflects the fact that an empty petri dish will remain empty for all time. The system also has $p = 1/2$ as a fixed point. This reflects the less obvious fact that for this model a half full dish of bacteria will remain half full for all time.

If we know the nature of a fixed point (attracting, repelling, or neither), we are able to reach important conclusions about many other initial conditions as well. In the previous example, $p = 0$ is a repelling fixed point. This means that no matter how small an amount of bacteria the petri dish initially contains, the amount of bacteria will grow. As another example of bacteria growth, let us consider the dynamical system $([0, 1], f)$ where f is graphed in Figure 2.1. The first four iterates of $x_0 = 3/10$, shown in the figure, seem to indicate that the fixed point $p = 0$ is attracting for $([0, 1], f)$. If it is attracting, then if too small a fraction of the dish is initially covered with bacteria, over time the bacteria will die out. The dramatically different outcomes for the same amount of bacteria in the two models illustrate the need to be able to definitively determine the nature of the fixed points of a dynamical system.

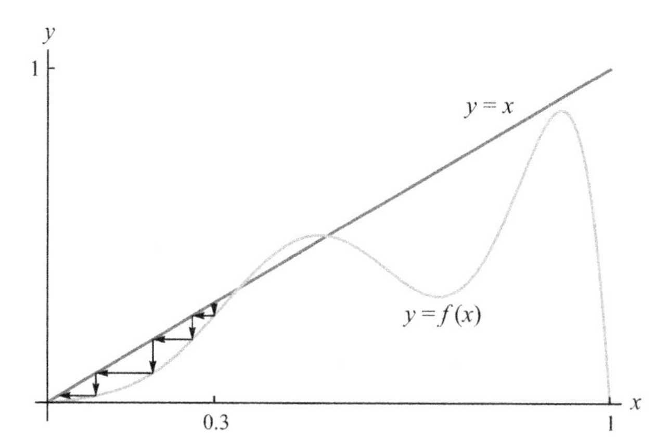

Figure 2.1. A web diagram illustrating the first four iterates under f of $x_0 = 3/10$.

When a fixed point p is attracting, we understand the long term behavior of all initial conditions in the basin of attraction of p. Both $([0, 1], g_2)$ and $([0, 1], f)$ have $p = 1/2$ as an attracting fixed point. In $([0, 1], g_2)$, $p = 1/2$ has $(0, 1)$ as its basin of attraction; thus, provided the petri dish is not empty or completely full, any initial amount of bacteria will, over time, cover about half the dish. This example might lead us to believe that a basin of attraction will always be an interval. Indeed, it follows from the definition of an attracting fixed point that its basin of attraction includes an interval centered around the fixed point. However, basins of attraction can be more complicated. This is illustrated by looking at some initial conditions for the dynamical system $([0, 1], f)$. The web diagram on the left in Figure 2.2 suggests that iterates of $x_0 = 0.73$ do not approach $p = 1/2$ and thus if the basin of attraction were an interval containing p, then it could not contain initial conditions greater than 0.73. However the web diagram on the right in Figure 2.2 suggests that $x_0 = 0.74$ is in the basin of attraction for p.

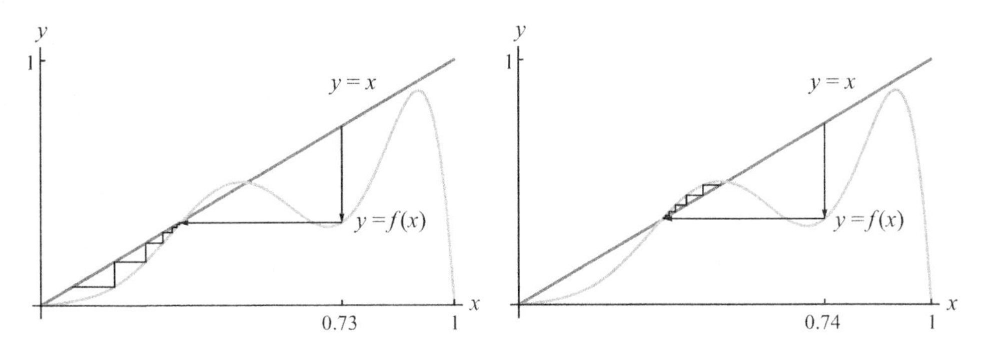

Figure 2.2. Web diagrams hinting at the complexity of basins of attraction.

The discussion above highlights two fundamental questions about a dynamical system: what is the nature of each of its fixed points and what is the basin of attraction of each attracting fixed point? The first question is the subject of the remainder of this module. Unfortunately, there is no easy answer to the second question. The project will show that determining the basin of attraction of an attracting fixed point can be nontrivial. In fact, in some cases it can be practically impossible to find.

In an attempt to find a general principle which will allow us to determine the nature of a fixed point, we turn our attention to the geometry of the graph near a fixed point. Consider the functions $g_{1/2}$ and g_2, both of which have a fixed point at $p = 0$. The web diagrams in Figure 2.3 suggest that $p = 0$ is an attracting fixed point for $g_{1/2}$ and a repelling fixed point for g_2.

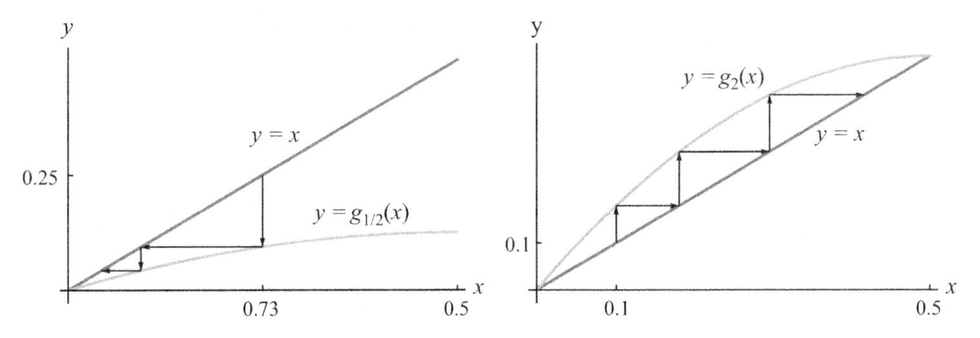

Figure 2.3. Web diagrams suggesting the nature of fixed point $p = 0$ for $g_{1/2}$ and g_2.

The only difference between the graphs of $g_{1/2}$ and g_2 at the origin is their steepness. To see how this affects the behavior of iterates of nearby initial conditions, consider the tangent line to the graph of $g_{1/2}$ at the origin. Its slope is $g'_{1/2}(0) = 1/2$. For x values near zero, the slope of the secant line approximates the slope of the tangent line, and

$$\frac{|g_{1/2}(x) - g_{1/2}(0)|}{|x - 0|} = \frac{|g_{1/2}(x)|}{|x|} < 1.$$

Thus the numerator $|g_{1/2}(x)|$ must be smaller than the denominator $|x|$ and $g_{1/2}(x)$ is closer to zero than x is to zero. In the case of g_2, $g'_2(0) = 2$ and a similar argument can be made to show that if x is an initial condition close to zero, then $g_2(x)$ is further away from zero than x.

These examples provide intuition for the ideas in the following theorem:

Theorem 2.1. *Let (X, f) be a dynamical system with fixed point p. If f is differentiable with a continuous derivative at p, then*

1. *if $|f'(p)| > 1$ then p is repelling,*
2. *if $|f'(p)| < 1$ then p is attracting, or*
3. *if $|f'(p)| = 1$ then p could be repelling, attracting, or neither.*

Before we discuss the proof of Theorem 2.1, we first need to examine the notion of continuity, as the continuity of the derivative of the function f will be critical in the proof. Intuitively, we think of continuous functions as those whose graphs can be drawn without lifting up our pencil. This understanding of continuity serves us well in calculus classes where we show that polynomials, exponential functions, and the sine and cosine functions are all continuous, but it is not precise enough for the work we will do in this book. Here we will need two characterizations of continuity. The first, given in Definition 2.2, is usually encountered in first-semester calculus and was used in Lemma 1.6. A second characterization is given by Proposition 2.3 and can be found in any undergraduate analysis textbook.

Definition 2.2. *The function $g : X \to Y$ is continuous if for all $a \in X$ and any sequence of points $\{x_n\}_{n=1}^{\infty}$ with $\lim_{n\to\infty} x_n = a$, we have $\lim_{n\to\infty} g(x_n) = g(a)$.*

Proposition 2.3. *A function $g : X \to Y$ is continuous if and only if for all $a \in X$ and $\epsilon > 0$, there exists $\delta > 0$ so that if $|x - a| < \delta$ then $|g(x) - g(a)| < \epsilon$.*

We will prove the first part of Theorem 2.1. The rest of the proof is completed in Exercises 2b and 7. We will prove Proposition 2.3 in Exercise 6.

Proof. Assume that $|f'(p)| > 1$. To show that p is a repelling fixed point, we must show that it satisfies Definition 1.4. Namely, we must find an open interval around p with the property that all initial conditions from this interval will leave the interval under iteration by f. We will do this by first finding an interval around p with the property that for all x in the interval, $|f'(x)|$ is a fixed amount larger than 1. We will then show that knowing this fact about the derivative means that points in the interval will leave the interval under iteration.

To find such an interval note that since the inequality is strict, there is a real number k such that

$$|f'(p)| > k > 1.$$

Let ϵ be the distance between k and $|f'(p)|$ and note that $\epsilon > 0$. By the continuity of f' and of the absolute value function, we can find a $\delta > 0$ such that for all $x \in (p - \delta, p + \delta)$,

$$\big|\, |f'(p)| - |f'(x)| \,\big| < \epsilon$$

and thus $|f'(x)| > k$.

We now argue that all initial conditions in the interval $(p - \delta, p + \delta)$ will eventually leave it under iteration by f. Choose $x \in (p - \delta, p + \delta)$ and note that by the mean value theorem there is an $\overline{x}$ between p and x such that

$$\frac{|f(x) - f(p)|}{|x - p|} = |f'(\overline{x})|.$$

Notice that $\overline{x}$ is also in $(p-\delta, p+\delta)$ and therefore by our choice of δ we have

$$|f'(\overline{x})| > k > 1.$$

Since p is a fixed point, putting these inequalities together we now have

$$\frac{|f(x)-f(p)|}{|x-p|} = \frac{|f(x)-p|}{|x-p|} = |f'(\overline{x})| > k > 1$$

and therefore

$$|f(x)-p| > k|x-p|.$$

This says that for initial conditions x in $(p-\delta, p+\delta)$, the distance between $f(x)$ and the fixed point p is greater than the distance between x and p. Either $f(x)$ is not in the interval $(p-\delta, p+\delta)$, or we can repeat the argument with x replaced by $f(x)$ to show that

$$|f^2(x)-p| > k|f(x)-p|.$$

Then, using the inequality obtained previously, we conclude that

$$|f^2(x)-p| > k^2|x-p|.$$

The argument can be repeated as long as $f^n(x)$ is in $(p-\delta, p+\delta)$ to give the inequality

$$|f^n(x)-p| > k^n|x-p|.$$

Since $k > 1$, there is an n such that $k^n|x-p| > \delta$ and $f^n(x)$ is not in the interval $(p-\delta, p+\delta)$. Thus, p is a repelling fixed point. □

We can now use Theorem 2.1 to analyze the fixed points of the dynamical systems $([0,1], g_c)$. For any c, the function g_c satisfies the hypotheses of the theorem and has the initial condition $p = 0$ as a fixed point. Since $g_c'(x) = c - 2cx$, we have $g_c'(0) = c$. Applying Theorem 2.1, we have that if $|c| > 1$ then $p = 0$ is repelling, and if $-1 < c < 1$ then it is attracting. The functions $g_{1/2}$ and g_2 provide examples. The theorem tells us nothing about the nature of the fixed point $p = 0$ for $g_1(x) = x(1-x)$ and $g_{-1}(x) = -x(1-x)$; further work is required for these cases (see Exercise 2a).

Analyzing the nature of the fixed points of any iterative process is an important application of Theorem 2.1. One example of an iterative process often familiar to calculus students is Newton's method, which is used to find approximations to roots of differentiable functions. The method approximates roots of a differentiable function f by iterating the function $N(x) = x - \frac{f(x)}{f'(x)}$. Note that x is in the domain of N if and only if $f'(x) \neq 0$.

If p is a root of f and $f'(p) \neq 0$, then it is a fixed point of N since $N(p) = p - \frac{0}{f'(p)}$. Conversely, if p is a fixed point of N, $p = p - \frac{f(p)}{f'(p)}$ and p must be a root of f. Thus the roots of f with nonzero derivatives are exactly the fixed points of N.

If a fixed point p is repelling, what are the implications for Newton's method? What are the implications if p is attracting? By iterating N, can one always be assured that the process will yield an approximation of an actual root of f? In the case of multiple roots, which root will the method approximate? These questions are explored in the project.

Exercises

1. Find and classify the fixed points of the functions $h(x) = \sqrt{x}$ and $f(x) = x^2$.

2. (a) Is the fixed point at $p = 0$ attracting, repelling, or neither for g_1 and g_{-1}? Explain your reasoning.

   (b) Give examples of three functions illustrating the three possibilities of (3) of Theorem 2.1, thus providing a proof of (3).

3. Sketch by hand the graph of a function with five fixed points, illustrating each of the possibilities described in Theorem 2.1.

4. The point $p = 0$ is a fixed point of $(\mathbb{R}, g_c)$ for all values of c.

   (a) For which values of c is there a second fixed point?

   (b) Use Theorem 2.1 to definitively answer the first question posed in the exploration.

5. In the proof of (1) of Theorem 2.1,

   (a) why is the real number k introduced?

   (b) why is the continuity of the derivative necessary?

   (c) why does f satisfy the hypotheses of the mean value theorem?

6. In this problem, we will explore the two definitions of continuity.

   (a) Let $f(x) = x^2$. Fix $\epsilon = 1/10$. Find a δ that satisfies Proposition 2.3 for

      i. $a = 1$

      ii. $a = 1/2$.

   (b) Prove Proposition 2.3.

7. Prove (2) of Theorem 2.1.

8. Let p be a fixed point of $N(x) = x - \frac{f(x)}{f'(x)}$, where f is a twice differentiable function and $f'(p) \neq 0$.

   (a) Show that p is an attracting fixed point.

   (b) What are the implications of the fact that p is attracting for Newton's method?

9. Consider a general second degree polynomial $f(x) = (x - a)(x - b)$. Assume $0 < a < b$.

   (a) Give a rough sketch of the graph of N (defined in Exercise 8) together with the line $y = x$. Use web diagrams to guess the basins of attraction for the attracting fixed points.

   (b) Prove your guess is correct. (Hint: A carefully drawn graph in part (a) and Theorem 1.5 and Lemma 1.6 might be helpful.)

Project

The function N used in Newton's method for finding approximations to the roots of $f(x) = x(x^2 - 4)$ is

$$N(x) = x - \frac{x^3 - 4x}{3x^2 - 4} = \frac{2x^3}{3x^2 - 4}.$$

Find the basin of attraction of each fixed point of the function N. What does the structure of the basin of attraction imply for Newton's method?

Describe the set E of initial values in the domain of N for which Newton's method will fail. Explain graphically why it fails in these cases.

This project is based on [Wa] where many other aspects of Newton's method applied to cubic polynomials are explored.

Module 3

Cycles and Their Classification

Exploration

Find all possible long term behaviors of initial conditions for the dynamical system $(\mathbb{R}, h)$ where $h(x) = x^2 - 1$.

Exposition

The goal of studying a dynamical system (X, f) is to gain an understanding of the long term behavior of all possible initial conditions. As we have seen, fixed points are examples of initial conditions whose long term behavior is easily understood; they do not change under iteration and their orbits consist of a single point. It is also easy to understand the long term behavior of an initial condition that lies in the basin of attraction of a fixed point p. Although its orbit may contain infinitely many elements, they form a sequence of points approaching p.

There are other situations where the long term behavior of an initial condition is easily understood. Imagine a frictionless, pocketless pool table where, once set in motion by the pool stick, a ball will continue to bounce about the table forever. Let the set X represent all possible ways in which a ball can hit the edge of the table; points in X are pairs of numbers specifying a location on the edge of the table and the angle with which the ball hits that location. Using the geometry of the table and the fact that the angle of incidence equals the angle of reflection, a ball's entire future trajectory is determined by the value $x \in X$ describing any one of its collisions with the edge. We can define a function f so that given a value x, $f(x) \in X$ describes the ball's subsequent collision with the edge.

If we assume that all balls have some velocity, the dynamical system (X, f) has no fixed points, and yet it has initial conditions whose long term behavior is simple. For example, consider a ball hit in such a way that it bounces back and forth between spots on opposite sides of the pool table. While this orbit has more than one element, it is, like the orbit of a fixed point, finite. This is an example of an orbit of a periodic point.

Definition 3.1. *Let $f : X \to X$ and $n \in \mathbb{N}$. A point $x \in X$ is a* **periodic point of period n** *for f if $f^n(x) = x$. If n is the smallest positive integer for which $f^n(x) = x$ we say x has* **minimal period n**. *The orbit of a periodic point with minimal period n is called an* **n-cycle**. *A point is called* **eventually periodic** *if one of its iterates is periodic.*

We say $x \in X$ is a periodic point of period n for the dynamical system (X, f) if it is a periodic point of period n for f.

Consider the dynamical system $([0, 1], g_{3.5})$. For notational convenience, set $g = g_{3.5}$. If we want to find the periodic points of period two, it is clear from Definition 3.1 that we need to find the points of intersection of the graph g^2 and the line $y = x$ as shown in Figure 3.1.

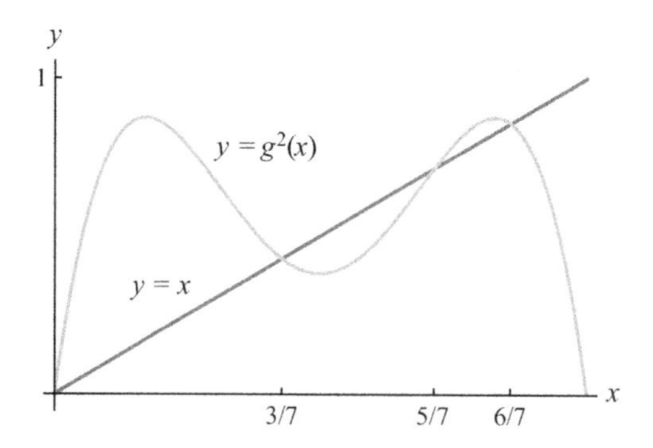

Figure 3.1. Graph of g^2 and the line $y = x$.

We solve the equation $g^2(x) = x$ to obtain the following points of period two for $([0, 1], g)$:

$$0, \frac{3}{7}, \frac{5}{7}, \frac{6}{7}.$$

When we iterate these points under g we see that

$$g(0) = 0,\ g\left(\frac{3}{7}\right) = \frac{6}{7},\ g\left(\frac{6}{7}\right) = \frac{3}{7},\ \text{and } g\left(\frac{5}{7}\right) = \frac{5}{7}.$$

Thus $\{3/7, 6/7\}$ is a 2-cycle, and $3/7$ and $6/7$ are each periodic points of minimal period two. The initial conditions 0 and $5/7$ are fixed points of $([0, 1], g)$, and therefore they are periodic points of period two but not of minimal period two. In fact, as fixed points they are periodic points of period k for any $k \in \mathbb{N}$. Similarly, $3/7$ and $6/7$ are periodic points of period $2k$ for any $k \in \mathbb{N}$. We see this for $3/7$ as follows:

$$g^{2k}\left(\frac{3}{7}\right) = g^{2(k-1)}\left(g^2\left(\frac{3}{7}\right)\right) = g^{2(k-1)}\left(\frac{3}{7}\right) = g^{2(k-2)}\left(g^2\left(\frac{3}{7}\right)\right) = \cdots = \frac{3}{7}.$$

This argument can be generalized to prove one direction of the following lemma. The complete proof is required in Exercise 3.

Lemma 3.2. *Suppose that an initial condition x has minimal period n. Then x is a periodic point of period m if and only if n divides m.*

Since x is a periodic point of period n for f exactly when it is a fixed point for f^n, we can classify the periodic points as attracting or repelling using Definitions 1.3 and 1.4 from Module 1.

Definition 3.3. *Let x be a periodic point of minimal period n for f. We say that x is an* **attracting periodic point** *for f if it is an attracting fixed point for f^n. We say that x is a* **repelling periodic point** *for f if it is a repelling fixed point for f^n.*

This brings up the following question: is it possible for one point in an n-cycle to be attracting for f^n while another is repelling? This question is answered for continuous functions by the following theorem.

Theorem 3.4. *Let $f : X \to X$ be a continuous function and let $\{p, f(p), \ldots, f^{n-1}(p)\}$ be an n-cycle. If any point in the n-cycle is repelling (attracting) for f^n, then all points in the n-cycle are repelling (attracting).*

Proof. We will prove that if $\{p, f(p)\}$ is a 2-cycle and one of the points in the 2-cycle is repelling for f^2, then both points in the 2-cycle are repelling. Exercise 8 requires the proof of the fact that if one point in the 2-cycle is attracting, then both points in the 2-cycle are attracting. The arguments for a general n-cycle are similar.

Let $f(p)$ be repelling for f^2. By Definition 1.4, this implies that there exists $\epsilon > 0$ such that if $y \in (f(p) - \epsilon, f(p) + \epsilon)$ then there exists $k > 0$ such that $(f^2)^k(y) \notin (f(p) - \epsilon, f(p) + \epsilon)$. We would like to show that p is also repelling for f^2.

First note that because f is continuous, there exists $\delta > 0$ such that for any x in $(p - \delta, p + \delta)$ we have $f(x) \in (f(p) - \epsilon, f(p) + \epsilon)$. We will show that p is repelling for f^2 by showing that for every $x \in (p - \delta, p + \delta)$, there exists an iterate of x under f^2 that does not lie in $(p - \delta, p + \delta)$.

Pick an x in $(p-\delta, p+\delta)$. By the choice of δ, $f(x)$ must lie in $(f(p)-\epsilon, f(p)+\epsilon)$. As stated above, this means there exists $k > 0$ such that $(f^2)^k(f(x)) \notin (f(p)-\epsilon, f(p)+\epsilon)$. We will show that $(f^2)^k(x) \notin (p-\delta, p+\delta)$ by contradiction.
Assume the opposite. That is, assume

$$(f^2)^k(x) \in (p-\delta, p+\delta).$$

By our definition of δ, we have

$$f\left((f^2)^k(x)\right) \in (f(p)-\epsilon, f(p)+\epsilon).$$

We can rewrite this as

$$(f^2)^k(f(x)) \in (f(p)-\epsilon, f(p)+\epsilon).$$

Yet k was chosen so that $(f^2)^k(f(x))$ would not lie in this interval, and we have reached a contradiction.

We conclude that $(f^2)^k(x) \notin (p-\delta, p+\delta)$, and we have shown that p is a repelling point for f^2. □

Since the nature of all the points in the orbit of a periodic point of period n is the same, it makes sense to talk about an attracting or repelling n-cycle. Thus we can now make the following definition.

Definition 3.5. *Let $\{p, f(p), \ldots, f^{n-1}(p)\}$ be an n-cycle for the dynamical system (X, f). We say it is a* **repelling n-cycle** *if p is a repelling periodic point for f, and it is an* **attracting n-cycle** *if p is an attracting periodic point for f.*

We know what it means graphically for a fixed point of f^n to be repelling (attracting), but what does it mean graphically for an n-cycle of f to be repelling (attracting)? We return to the dynamical system $([0,1], g)$ discussed earlier. Figure 3.2 shows the 2-cycle $\{3/7, 6/7\}$ in the graph on the left and iterates of the point $x_0 = 1/3$ in the graph on the right. It appears that the iterates of x_0 are keeping a fixed distance away from the 2-cycle, leading us to conjecture that the 2-cycle $\{3/7, 6/7\}$ is repelling.

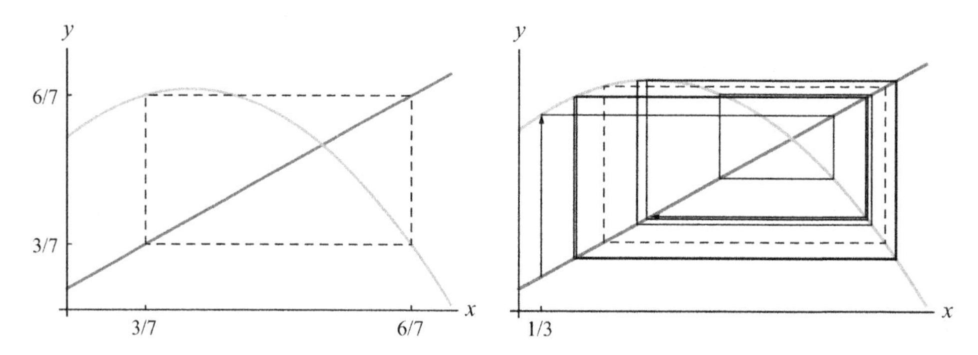

Figure 3.2. The 2-cycle $\{3/7, 6/7\}$ (dashed lines) and iterates of $x_0 = 1/3$.

On the other hand, when we consider the dynamical system $([0,1], g_{3.4})$ we observe a different sort of behavior. This dynamical system has a 2-cycle that we will denote by $\{p, g_{3.4}(p)\}$. Figure 3.3 shows this 2-cycle in the graph on the left and iterates of the point

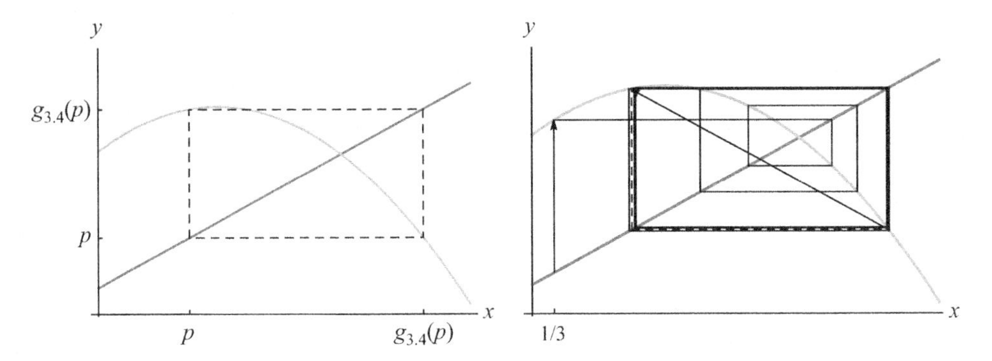

Figure 3.3. The 2-cycle $\{p, g_{3.4}(p)\}$ (dashed lines) and iterates of $x_0 = 1/3$.

$x_0 = 1/3$ in the graph on the right. In this case the iterates of x_0 seem to approach the elements of the 2-cycle. The orbit of x_0 begins to look like a period two orbit in the sense that elements of the orbit oscillate between values approaching p and $g_{3.4}(p)$. It appears that in this case the 2-cycle $\{p, g_{3.4}(p)\}$ is attracting.

Figure 3.2 is consistent with what we saw in the proof of Theorem 3.4. There we found that for each x close to p, there was an iterate of f^2 that moved both x away from p and $f(x)$ away from $f(p)$. Similar scrutiny of the argument used in Exercise 8 will show it to be consistent with the behavior observed in Figure 3.3.

Finally we point out that if f and its first n derivatives are continuous, we can use Theorem 2.1 to easily determine the nature of an n-cycle. For example, provided the value of $\left|\frac{d}{dx}\left[g^2(3/7)\right]\right|$ is not 1, we can determine if $3/7$ is an attracting or repelling periodic point for $([0,1], g)$. We calculate the derivative of g^2 using the chain rule:

$$\left|\frac{d}{dx}\left[g \circ g(x)\right]\right| = |g'(g(x)) \cdot g'(x)|.$$

Thus

$$\begin{aligned}
\left|\frac{d}{dx}\left[g \circ g\left(\frac{3}{7}\right)\right]\right| &= \left|g'\left(g\left(\frac{3}{7}\right)\right) \cdot g'\left(\frac{3}{7}\right)\right| \\
&= \left|g'\left(\frac{6}{7}\right) \cdot g'\left(\frac{3}{7}\right)\right| \\
&= \left|\left(\frac{1}{2}\right)\left(-\frac{5}{2}\right)\right| > 1
\end{aligned}$$

and, as we conjectured from the graphs in Figure 3.2, $x = 3/7$ is a repelling periodic point for g. When we use the chain rule to calculate the derivative of g^2 at $6/7$, the other point in the orbit of $3/7$, we see that

$$\begin{aligned}
\left|\frac{d}{dx}\left[g \circ g\left(\frac{6}{7}\right)\right]\right| &= \left|g'\left(\frac{3}{7}\right) \cdot g'\left(\frac{6}{7}\right)\right| \\
&= \left|\frac{d}{dx}\left[g \circ g\left(\frac{3}{7}\right)\right]\right|.
\end{aligned}$$

The derivative of g^2 at $6/7$ is the same as the derivative of g^2 at $3/7$, showing clearly that the nature of both points in the 2-cycle is the same.

In general, for a continuous function f whose first n derivatives are continuous, the value of the derivative of f^n will be the same for all points in an n-cycle. Thus the derivative at any point in the cycle can be used to determine the nature of all points in the cycle. To see this, suppose x_0 has minimal period n for f. Let $x_i = f^i(x_0)$, $1 \leq i \leq n-1$. Then $\{x_0, x_1, x_2, \ldots, x_{n-2}, x_{n-1}\}$ is an n-cycle. Using repeated applications of the chain rule (Exercise 6) we see that

$$\frac{d}{dx}[f^n(x_0)] = f'(x_{n-1})f'(x_{n-2})\cdots f'(x_1)f'(x_0).$$

The derivative of f^n evaluated at a point in an n-cycle is the product of the derivatives of f evaluated at each of the points in the n-cycle. Since the choice of which point in the n-cycle is called x_0 is arbitrary, the derivative of f^n is the same for any point in the cycle.

We have seen how important attracting fixed points and attracting periodic points are to the long term behavior of initial values in a dynamical system, as they determine the long term behavior of the values in their basins of attraction. But repelling fixed points and repelling periodic points can also be important for the long term behavior of many initial values in our phase space, as you will see in the project.

Exercises

1. Consider the dynamical system (X, g_4) where $X = [0, 1]$ and $g_4(x) = 4x(1-x)$.

   (a) Approximate all period 4 points.

   (b) List all minimal period 4 points.

   (c) List all 4-cycles.

   (d) Are the 4-cycles attracting, repelling, or neither?

2. (a) Suppose that a point x is periodic with period 12. What are the possible minimal periods of x?

   (b) The graph in Figure 3.4 shows five points of intersection of h^3 and $y = x$ for some continuous function h. Is it necessarily the case that the dynamical

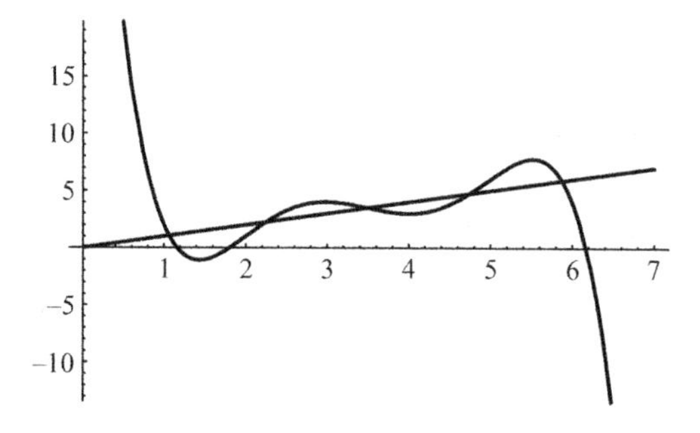

Figure 3.4. Graph of h^3 for Exercise 2.

system $(\mathbb{R}, h)$ has one or more 3-cycles? Is it possibly the case that the dynamical system $(\mathbb{R}, h)$ has one or more 3-cycles? List and explain the various possibilities.

3. Let $f : X \to X$ and $n \in \mathbb{N}$.

   (a) Generalize the argument provided in the exposition to show that if $x \in X$ has minimal period n, then it is a periodic point of period nk for all $k \in \mathbb{N}$.

   (b) Show that if $x \in X$ is periodic with period m and minimal period n, then n divides m.

4. (a) Give an example of a function defined on $\mathbb{R}$ for which every orbit is infinite.

   (b) Give an example of a function defined on $[0, 1]$ for which every orbit consists of a single fixed point.

   (c) Give an example of a function defined on $[0, 1)$ for which every orbit is a 2-cycle. (Hint: Try a piecewise defined function.)

5. In this module, you have learned what it means for a function to have a periodic point. In your previous mathematics classes you have encountered periodic functions. In this exercise we compare these two notions of periodicity.

   (a) If h is a periodic function does this mean that all initial conditions are periodic? If so, explain why. If not, give a counterexample.

   (b) If all initial conditions are periodic, is h a periodic function? If so, explain why. If not, give a counterexample.

6. In this exercise, we verify that the derivative of f^n is the same for all points in an n-cycle and thus they all have the same nature as fixed points of f^n.

   (a) Use the chain rule to show that if $\{x_0, x_1, x_2\}$ is a three cycle then

$$\frac{d}{dx}\left[f^3(x_0)\right] = f'(x_0)f'(x_1)f'(x_2).$$

   (b) If $\{x_0, x_1, x_2, \ldots, x_{n-1}\}$ is an n-cycle, use the chain rule and an induction argument to show that

$$\frac{d}{dx}[f^n(x_0)] = f'(x_{n-1})f'(x_{n-2})\cdots f'(x_1)f'(x_0).$$

7. Let $f : X \to X$ be a continuous function and let $\{p, f(p), \cdots, f^{n-1}(p)\}$ be an attracting n-cycle. Let x be an element in its basin of attraction. Does $\lim_{k\to\infty} f^k(x)$ exist? If so, explain why. If not, give a counterexample.

8. Let $f : X \to X$ be a continuous function and let $\{p, f(p)\}$ be a 2-cycle. Without using Theorem 2.1, prove that if one of the points in the 2-cycle is attracting for f^2, then the other is too.

Project

Define f on $[0,1]$ by

$$f(x)=\begin{cases}2x & 0\le x\le \frac{1}{2}\\ 2x-1 & \frac{1}{2}<x\le 1.\end{cases}$$

Consider the dynamical system $([0,1], f)$. Give a complete description of the periodic points and the eventually periodic points, and discuss the long term behavior of initial values in this system.

You will next use your understanding of $([0,1], f)$ to prove Fermat's[1] little theorem, which is an important result from number theory. It states that for all prime numbers p and all integers a,

$$a^p - a \equiv 0 \mod p.$$

Prove the special case of this theorem when $a=2$ by finding the relationship between the quantity 2^p-2 and the number of periodic points of period p of the dynamical system $([0,1], f)$.

Prove the statement for the special case when $a=3$ similarly by finding an appropriate dynamical system.

Prove the general statement using the methods you discovered above.

This project is based on [FJS].

[1] Pierre de Fermat (1601–1665) was a French lawyer and a mathematician. He first stated this theorem in a letter in 1640. He also developed a method to determine maxima and minima of various curves and is the first person known to have evaluated the integral of general power functions. His most famous theorem states that no three positive integers a, b, and c can satisfy the equation $a^n+b^n=c^n$ for any integer $n>2$.

Module 4

Bounded Orbits and Fractal Dimension

Exploration

Let

$$h(x) = \begin{cases} 4x & \text{if } x \le \frac{1}{2} \\ 4 - 4x & \text{if } x > \frac{1}{2} \end{cases}$$

and consider the dynamical system $(\mathbb{R}, h)$. Find the set C of initial conditions whose orbits remain in the interval $[0, 1]$. Can you describe the behaviors of orbits of initial conditions in C?

Exposition

When a dynamical system models a physical situation, there is often some bounded range of initial conditions that are realistic in the context of that physical situation. The study of the behavior of orbits that remain in this range is then of particular importance since they represent the only possible realistic long term scenarios. In Module 1, for example, we studied the dynamical system $(\mathbb{R}, g)$ where $g(x) = 2x(1-x)$, and we noticed that the orbit of any initial condition from $[0, 1]$ stays in $[0, 1]$ under iteration. Thus, when using the function g to model the growth of bacteria in a petri dish, we were able to define a new dynamical system (X, g) where $X = [0, 1]$ represented the possible proportions of the dish covered by bacteria.

An orbit that remains within a bounded interval is called a **bounded orbit**. Our work in Module 1 led to a complete understanding of the bounded orbits of the dynamical system (X, g). There are three finite bounded orbits: the orbits of the two fixed points, each of which consists of a single point, and the orbit of the eventually fixed point $x = 1$ which consists of two points. Every other initial condition in $[0, 1]$ has an infinite bounded orbit with iterates approaching the attracting fixed point $p = 1/2$.

A point has a finite orbit if and only if it is either periodic or eventually periodic (Exercise 1), and finite orbits are always bounded. We will focus our attention on understanding infinite, bounded orbits. The infinite, bounded orbits in (X, g) exhibited only a single type of long term behavior, and in this module we will ask whether that behavior is typical. That is, do all infinite, bounded orbits approach an attracting periodic orbit?

To investigate this question, let us consider the dynamical system $(\mathbb{R}, f)$ where

$$f(x) = \begin{cases} 3x & \text{if } x \leq \frac{1}{2} \\ 3 - 3x & \text{if } x > \frac{1}{2}. \end{cases}$$

This function is particularly useful for such an investigation because all its periodic orbits are repelling (Exercise 4). Thus if an infinite, bounded orbit exists, it must exhibit a new sort of behavior. We note that the iterates of an initial condition outside of the interval $[0, 1]$ approach negative infinity (Exercise 2). We thus focus our attention on the set of initial conditions from $[0, 1]$ whose orbits remain in $[0, 1]$. We call this set Γ.

Let Γ_1 be the set of initial conditions that remain in $[0, 1]$ after one iteration. Clearly Γ is a proper subset of Γ_1, since there are initial conditions, for example $x_0 = 1/6$, for which $f(x_0) \in [0, 1]$ but $f^2(x_0) \notin [0, 1]$. Since $f([0, 1/3]) = f([2/3, 1]) = [0, 1]$, we have that

$$\Gamma_1 = \left[0, \frac{1}{3}\right] \cup \left[\frac{2}{3}, 1\right].$$

This is illustrated in Figure 4.1.

Let Γ_2 be the set of initial conditions that remain in $[0, 1]$ after two iterations of f. It is clear that $\Gamma_2 \subseteq \Gamma_1$. To find Γ_2 we need to identify those initial conditions from $[0, 1]$ that are in $\Gamma_1 = [0, 1/3] \cup [2/3, 1]$ after one iteration. We see that $f(1/9) = f(8/9) = 1/3$ and $f(2/9) = f(7/9) = 2/3$. Thus, as illustrated in Figure 4.2,

$$\Gamma_2 = \left[0, \frac{1}{9}\right] \cup \left[\frac{2}{9}, \frac{1}{3}\right] \cup \left[\frac{2}{3}, \frac{7}{9}\right] \cup \left[\frac{8}{9}, 1\right].$$

We can continue in this way to define for any arbitrary integer $n > 0$ the set

$$\Gamma_n = \{x \mid 0 \leq f^i(x) \leq 1 \text{ for } i = 0, 1, \cdots, n\}.$$

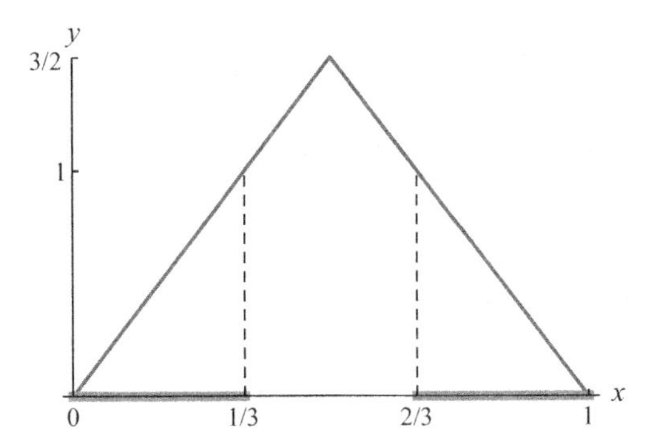

Figure 4.1. Initial conditions x for which $f(x)$ is in $[0, 1]$.

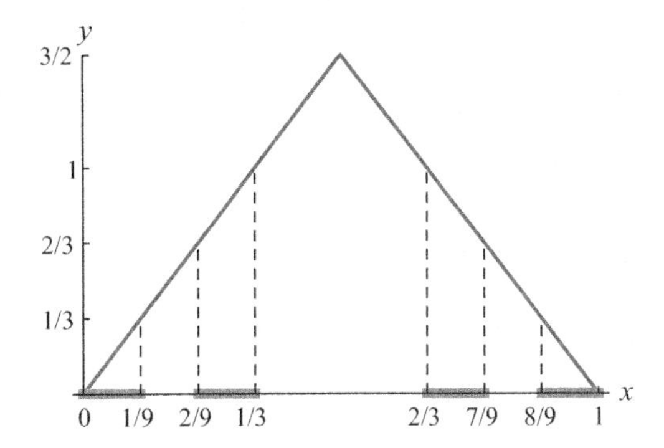

Figure 4.2. Initial conditions x for which $f(x)$ is in Γ_1.

This is the set of initial conditions in $[0, 1]$ that stay in $[0, 1]$ for the first n iterates, and as before $\Gamma \subseteq \Gamma_n \subseteq \Gamma_{n-1}$. The set Γ_n consists of 2^n intervals, each with length $1/3^n$ and of the form $[k/3^n, (k+1)/3^n]$ (Exercise 3b).

Since x is in Γ if and only if it is in Γ_n for all $n \in \mathbb{N}$, we have

$$\Gamma = \bigcap_{n \in \mathbb{N}} \Gamma_n.$$

The set Γ is called the Cantor set, named after the German mathematician Georg Cantor.[1]

Clearly there are many initial conditions in Γ with finite orbits. For example, the end-points of the intervals in Γ_n map to zero or one under f^n and so they are eventually fixed points. Thus they are in Γ and have finite orbits.

We now return to the question of whether there are any initial conditions in Γ with infinite orbits. We will not try to find such an initial condition explicitly. Instead, we will prove the existence of initial conditions with infinite orbits by an indirect argument. We will argue that the set of initial conditions with finite orbits is much smaller, in a precise sense, than the set Γ. In order to make this argument, we will need the following definitions and theorems. The proofs of Theorems 4.2 and 4.3 can be found in any real analysis or set theory text. A proof of Theorem 4.4 is outlined in Exercise 5.

[1]Georg Cantor (1845-1918) is often called the creator of modern set theory. He showed, among other things, that the rational numbers are countable while the real numbers are not. His groundbreaking work was vigorously opposed by some of his contemporaries.

Definition 4.1. *A set S is* **countable** *if it is finite or if there is a one-to-one, onto map $\phi : S \to \mathbb{N}$. A set is* **uncountable** *if it is not countable.*

Theorem 4.2. *The set of rational numbers is countable.*

Theorem 4.3. *Any subset of a countable set is countable.*

Theorem 4.4. *The Cantor set Γ is an uncountable set.*

We will now show that the set of initial conditions in Γ with finite orbits under f is countable by showing that it is a subset of the rational numbers. Then, since Theorem 4.4 establishes that the set Γ is uncountable, there must be points in Γ other than those with finite orbits.

We begin by investigating the periodic points. The fixed points are given by the intersection of the graph of f with the graph of the line $y = x$, and we can solve for the fixed points by solving the equations $3x = x$ and $3 - 3x = x$.

To find the period two points, we solve for the 2^2 intersections of the graph of f^2 with the graph of the line $y = x$. The function f^2 is given by

$$f^2(x) = \begin{cases} 3^2x & \text{if } x \leq \frac{1}{6} \\ 3 - 3^2x & \text{if } \frac{1}{6} < x \leq \frac{1}{2} \\ -6 + 3^2x & \text{if } \frac{1}{2} < x \leq \frac{5}{6} \\ 9 - 3^2x & \text{if } \frac{5}{6} < x \end{cases}$$

and the graph of f^2 and $y = x$ is shown in Figure 4.3. We can solve for the period two points by solving the equations $3x^2 = x$, $3 - 3^2x = x$, $-6 + 3^2x = x$, and $9 - 3^2x = x$.

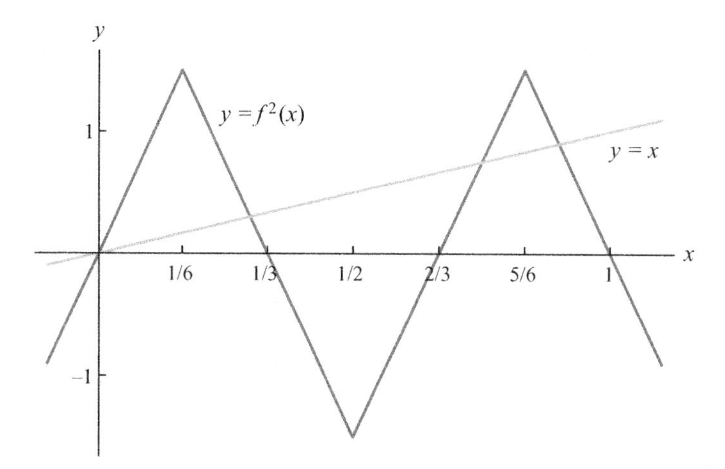

Figure 4.3. A graph of $y = f^2(x)$ and $y = x$.

The equations of the four lines making up the graph of f^2 have the form $y = a \pm 3^2x$ where a is an integer. The graph of f^3 will consist of eight line segments. Since f^3 is obtained by composing f with f^2, the equations of the eight lines have the form $y = a \pm 3^3x$. A simple induction argument shows that the equations giving the 2^n lines in the graph of f^n will have the form $y = a \pm 3^n x$. The period n points are solutions to equations of the form $a \pm 3^n x = x$ and, as such, are rational numbers.

Since it is clear that $f(x)$ can be rational if and only if x is rational, the eventually periodic points must also be rational. Thus the set of initial conditions in Γ that have a finite orbit is a subset of the rational numbers and so is countable by Theorems 4.2 and 4.3. Because Γ is uncountable and the set of initial conditions with a finite orbit is only a countable subset of Γ, there must be initial conditions in Γ that do not have finite orbits. Their orbits can not approach a periodic orbit since all periodic orbits for this dynamical system are repelling.

In iterating points for this dynamical system and for others, we have found points whose iterates appear to bounce about with no particular pattern. Until now, however, we could not be certain these points exhibited a new qualitative behavior. It is entirely possible that points whose iterates appear to bounce about are actually approaching an attracting periodic orbit of extremely large period. This example proves the existence of initial conditions with bounded infinite orbits that do not approach a periodic orbit.

The set Γ has many interesting properties. For instance, it is an example of a fractal set, a notion we will explore in more detail in Modules 10 and 12. We have also seen here that Γ is a surprisingly large set from the point of view of the number of points it contains. In the project you will investigate the size of Γ from a different point of view, namely that of dimension.

Exercises

1. Show that a finite orbit must be either periodic or eventually periodic.

2. Show that for the dynamical system $(\mathbb{R}, f)$, the iterates of any initial condition in $(-\infty, 0) \cup (1, \infty)$ tend towards negative infinity.

3. Create the graph of f^n for several values of $n \in \mathbb{N}$ and use them to provide a graphical justification of the following facts.

   (a) $(\mathbb{R}, f)$ has exactly 2^n periodic points of period n for all $n \in \mathbb{N}$.

   (b) Γ_n consists of 2^n intervals, each with length $1/3^n$ and of the form $[k/3^n, (k+1)/3^n]$.

   (c) Every initial condition in (Γ, f) has a periodic point within a distance of $1/3^n$ for each $n \in \mathbb{N}$.

4. Prove that all the periodic points of $(\mathbb{R}, f)$ are repelling.

5. The goal of this exercise is to prove that Γ is an uncountable set.

   (a) Every number in $[0, 1]$ can be expressed in terms of its base three, or **ternary**, expansion. That is, a point x can be written $.x_1x_2x_3\cdots$ where $x_i \in \{0, 1, 2\}$ and

   $$x = x_1 \cdot \frac{1}{3} + x_2 \cdot \frac{1}{3^2} + x_3 \cdot \frac{1}{3^3} + \cdots .$$

   Find a ternary expansion for $x = 1/3, 4/9, 26/27$, and $1/4$.

   (b) What must be true about the ternary expansion of points in Γ_1? In Γ_n? In Γ?

   (c) Use the ternary expansion of points in Γ and a diagonalization argument to show that Γ is uncountable.

6. In this exercise you will show three important properties of the set Γ.

   (a) A set $S \subset \mathbb{R}$ is **totally disconnected** if it contains no open intervals. Find an expression for the length of Γ_n for any $n \in \mathbb{N}$. Use it to show that Γ is totally disconnected.

   (b) A set $S \subset \mathbb{R}$ is **open** if every point in S is contained in an open interval I with $I \subseteq S$. A set is **closed** if its complement is open. Show that Γ is a closed set.

   (c) A set $S \subset \mathbb{R}$ is **perfect** if for every $x \in S$ and every interval I containing x, there is another point from S contained in I. Show that if x is in Γ, then there are other points in Γ within a distance of $1/3^n$ to x for any value of $n \in \mathbb{N}$ and conclude that Γ is a perfect set.

7. Does the set C from the exploration have the same properties (as given in Exercise 6) as the set Γ?

Project

Consider covering a set $S \subseteq \mathbb{R}$ by intervals of length $1/3$: how many such intervals do you need? If S is the unit interval, you could use exactly three such intervals to complete this task. However, if S is the Cantor set you only need two such intervals, since $\Gamma \subseteq [0, 1/3] \cup [2/3, 1]$. This gives us some indication of how these two sets are different. We can explore this difference further by considering intervals of length $1/9$. We see that we would now need at least nine such subintervals to cover $[0, 1]$ but only four to cover Γ. Continuing in this vein, we could make a similar comparison using intervals of any length ϵ. For $\epsilon > 0$, about $1/\epsilon$ are needed to cover $[0, 1]$ and less than that are needed to cover Γ. These ideas motivate the following definition.

Definition 4.5. *Fix a set $S \subset \mathbb{R}$. For $\epsilon > 0$, let $N(\epsilon)$ be the smallest number of intervals $\{I_i\}_{i=1}^{N(\epsilon)}$ of length ϵ such that $S \subseteq \bigcup_{i=1}^{N(\epsilon)} I_i$. The* **box-counting dimension of S**, *denoted dim S, is*

$$dim\ S = \lim_{\epsilon \to 0} \frac{\ln N(\epsilon)}{\ln(\frac{1}{\epsilon})}.$$

To familiarize yourself with this definition, find the box-counting dimension of

1. the interval $[a, b] \subseteq \mathbb{R}$ for $a < b$
2. a finite set of points
3. the Cantor set Γ
4. the set C from the exploration.

The last two sets have the following characteristic:

Definition 4.6. *A set S is said to have* **fractal dimension** *if dim S is not an integer.*

For what values of a and b can you find a dynamical system whose set of bounded orbits has fractal dimension $(\ln a)/(\ln b)$?

Module 5

Sensitive Dependence and Chaos

Exploration

Suppose a function is used to model some physical situation, for example the fraction of a petri dish covered by bacteria as time passes. In real life we can never measure initial conditions exactly, no matter how many decimal places of accuracy we use. If this is the case, in order to have confidence in the long term predictions of a mathematical model, the model must predict similar long term behavior for nearby initial conditions.
Define a function h_c by

$$h_c(x) = \begin{cases} cx & \text{if } x \leq \frac{1}{4} \\ \frac{c}{3} - \frac{c}{3}x & \text{if } x > \frac{1}{4}. \end{cases}$$

For what values of $1 \leq c \leq 4$ would you have confidence in the model given by $([0, 1], h_c)$?

Exposition

We have all had the frustrating experience of planning a weekend outing based on a prediction of beautiful weather only to have it rain when the weekend arrives. The mathematical models that meteorologists use to predict the weather are examples of dynamical systems. The meteorologist observes the initial condition x_0, perhaps the current temperature, pressure, and wind readings, and then iterates a function f using this initial condition. The iterate $f(x_0)$ gives the predicted weather one time unit into the future, $f^2(x_0)$ gives the predicted weather two time units into the future, and so on.

Unfortunately for our ability to plan weekend outings, the further the weather is predicted into the future using meteorological models, the less accurate the predictions become. You might think this is due entirely to the fact that weather is a complicated phenomenon and it is impossible to incorporate all the variables affecting the weather into a model. In fact, this is only part of the problem. Researchers have shown that even if all the factors influencing the weather could be included in the model, there is a limit on how far into the future the model will be able to predict due to our inability to measure initial conditions exactly. If the measured initial conditions are off by even the smallest amount from the true initial conditions, then the actual weather will diverge from the predicted weather.

This is an example of sensitive dependence on initial conditions, a phenomenon first described in the early 1960s by the meteorologist Edward Lorenz[1]. A function f has sensitive dependence on initial conditions if the iterates of a point x eventually separate from the iterates of other points that are arbitrarily close to x. The following definition is a formalization of this idea.

Definition 5.1. *Let (X, f) be a dynamical system. We say that f* **has sensitive dependence on initial conditions** *if there exists $D > 0$ such that for any $x \in X$ and any $\delta > 0$, there is $y \in X$ with $|y - x| < \delta$ and an $n \in \mathbb{N}$ such that $|f^n(x) - f^n(y)| > D$.*

We say (X, f) has sensitive dependence on initial conditions if f has sensitive dependence on initial conditions.

From a practical point of view, if a function has sensitive dependence on initial conditions then long term prediction will be impossible. Any slight difference in initial conditions could result in dramatically different long term behavior. This is a serious problem if the actual initial condition can only be measured approximately. Figure 5.1 illustrates this phenomenon: it shows the first ten iterates of 0.1 and 0.101 under the function h_4 defined in the exploration. As can be seen in the figure, if the actual initial condition was 0.101 but the measured initial condition was 0.1, then the model would be accurate only for the first five or six iterations. After that, the behavior of the iterates of 0.101 diverges significantly from the behavior of the iterates of 0.1.

The following theorem shows that this is not an isolated occurence.

Theorem 5.2. *The dynamical system $([0, 1], h_4)$ has sensitive dependence on initial conditions.*

The proof of Theorem 5.2 will rely on the fact that the image of an interval in $[0, 1]$ will spread out under iterations of h_4 to eventually include all of $[0, 1]$. The proof of this

[1] Edward Lorenz (1917-2008) held an undergraduate degree in mathematics and advanced degrees in both mathematics and meteorology. While a professor of meteorology at the Massachusetts Institute of Technology, Dr. Lorenz first noticed the phenomenon of sensitive dependence when he used slightly different initial conditions in his weather model and the model predicted vastly different future behavior.

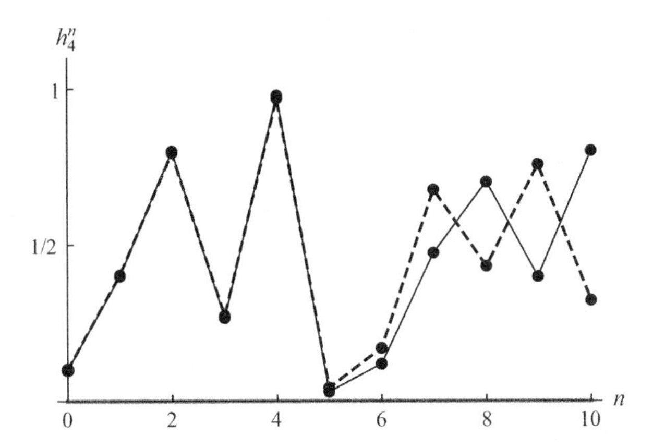

Figure 5.1. The first ten iterates under h_4 of $x_0 = 0.1$ (connected with a solid line) and $x_0 = 0.101$ (connected with a dashed line).

is the subject of Lemmas 5.3 and 5.4. We include the proof of Lemma 5.4; the proof of Lemma 5.3 is the subject of Exercise 9. In what follows, we will refer to h_4 as h for ease of notation.

Lemma 5.3. *For any b with $0 < b \leq 1$, there is an $n \in \mathbb{N}$ such that $h^n([0, b]) = [0, 1]$.*

Lemma 5.4. *For any a, b with $0 < a < b \leq 1$, there is an $n \in \mathbb{N}$ such that $h^n([a, b]) = [0, 1]$.*

Proof. First observe that $h([a, 1]) = [0, d]$ for some $0 < d \leq 1$ and the result now follows from Lemma 5.3. Next note that if $1/4$ is in $[a, b]$ then $h([a, b]) = [c, 1]$ for some $0 \leq c < 1$ and the result follows from our first observation. Thus we need only consider intervals $[a, b]$ that do not contain $1/4$ or 1. There are two cases: either $[a, b] \subset (0, 1/4)$ or $[a, b] \subset (1/4, 1)$.

Consider first the case of $[a, b] \subset (0, 1/4)$. Then $h(a) = 4a$ and $h(b) = 4b$. If $h(a) \leq 1/4 \leq h(b)$, then $1/4$ is in $h([a, b])$ and the result follows from the previous paragraph. Otherwise $h(a)$ and $h(b)$ fall on the same side of $1/4$, $h([a, b]) = [4a, 4b]$, and the length of $h([a, b])$ is four times the length of $[a, b]$. Continuing in this way, we see that either $h^k([a, b])$ contains $1/4$, in which case we are done, or the length of $h^k([a, b])$ is at least $(4/3)^k$ times the length of $[a, b]$. Since $h^k([a, b])$ is contained in $[0, 1]$, its length cannot increase indefinitely and eventually $h^k([a, b])$ must contain $1/4$ as desired. A similar argument holds for the second case. □

Theorem 5.2 is an immediate corollary of Lemma 5.4: given $x \in [0, 1]$ and a small interval containing x, the set of images of points in that interval will eventually contain all of $[0, 1]$. Thus for any $0 < D < 1/2$, the iterates of x will eventually be at least a distance D from the iterates of some other point in that interval.

Lemma 5.4 also shows that the system $([0, 1], h)$ satisfies another important property related to sensitive dependence on initial conditions.

Definition 5.5. *Let (X, f) be a dynamical system. We say that f is* **transitive** *if given any two nonempty, open intervals U, $W \subseteq X$, there exists $n \in \mathbb{N}$ such that $f^n(U)$ intersects W.*

We say the dynamical system (X, f) is transitive if f is transitive.

For the function h, given a pair of nonempty, open intervals U and W, we can choose n to be the iterate for which $h^n(U) = [0, 1]$. Clearly this iterate of U intersects W and thus $([0, 1], h)$ is an example of a transitive dynamical system.

Informally, a function f is transitive if every open interval visits every other open interval under iteration. Intuitively we can think of a transitive function as one that mixes things up. One manifestation of this property can be seen in the first 100 iterates under h of $x_0 = 1/3$. As shown in the web diagram in Figure 5.2, these iterates seem to bounce all about the phase space.

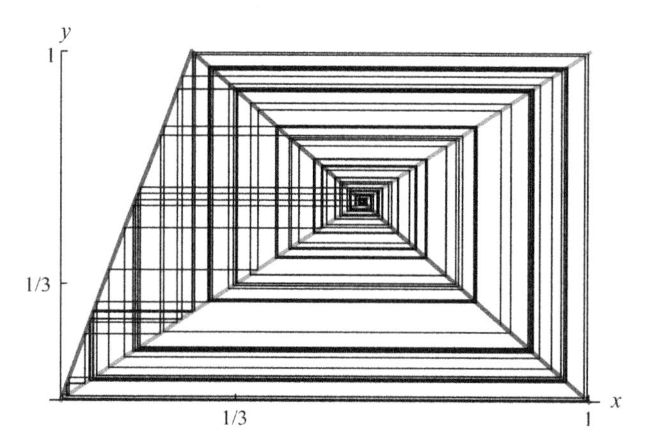

Figure 5.2. The first hundred iterates under h of $x_0 = 1/3$.

The structure of the set of periodic points of h is also interesting.

Definition 5.6. *Let $X \subseteq \mathbb{R}$. We say that $S \subseteq X$ is* **dense** *in X if given any $x \in X$ and any $\epsilon > 0$, there is a point $s \in S$ with $|x - s| < \epsilon$.*

Exercise 7 will verify that the periodic points of h form a dense subset of $[0, 1]$.

This combination of traits makes h an interesting function. First, the fact that it has sensitive dependence on initial conditions makes long term behavior unpredictable. On the other hand, there is regularity since the periodic points, whose long term behavior is completely understood, are ubiquitous. It is also transitive, which tells us that it has an element of indecomposability: since every open interval visits every other open interval under iteration, there is no way to divide the system into smaller pieces. These three properties form the basis for the definition of a chaotic dynamical system.

Definition 5.7. *Let (X, f) be a dynamical system. We say that f is* **chaotic** *if it is transitive, has sensitive dependence on initial conditions, and has a dense set of periodic points.*

We say the dynamical system (X, f) is chaotic if f is chaotic.

The system $([0, 1], h)$ is not unique, as there are many examples of chaotic dynamical systems. In particular, if X is an interval (not necessarily bounded) and f is continuous and transitive then (X, f) is chaotic. We refer the interested reader to [VB] for a proof.

The definition of chaos is also not unique, and a discussion of the various definitions used in the literature can be found in [MDS].

We close by making an important observation. In common speech, the adjective "chaotic" implies disorder and randomness. There is nothing random about a chaotic dynamical system: given an initial condition with complete precision, its orbit is determined. However, as we have seen, if f has sensitive dependence on initial conditions then our inability to measure our initial conditions exactly, and to compute with complete precision, introduces an element of uncertainty into our predictions. This is consistent with our informal understanding of chaos. The exercises and project will explore the three properties of chaos in more depth.

Exercises

1. By iterating points and their nearby neighbors, form a conjecture as to whether $g_c : [0, 1] \to [0, 1]$ has sensitive dependence on initial conditions for $2 < c \leq 4$. Recall that $g_c(x) = cx(1 - x)$.

2. Does the function $f : \mathbb{R} \to \mathbb{R}$ defined by $f(x) = 2x$ have sensitive dependence on initial conditions? Prove your assertion.

3. Does the function $f : \mathbb{R} \to \mathbb{R}$ defined by $f(x) = -x$ have sensitive dependence on initial conditions? Prove your assertion.

4. The image of the open interval $(-1, 1)$ under iterates of $f(x) = 2x$ eventually intersects any nonempty subset of $\mathbb{R}$. Is $f : \mathbb{R} \to \mathbb{R}$ transitive? Prove your assertion.

5. For $([0, 1], h_c)$ as defined in the exploration, determine whether h_c is transitive for $1 \leq c < 4$. Is it chaotic?

6. Are the following sets dense subsets of $[0, 1]$? Prove your assertions.
   (a) the rational numbers in $[0, 1]$
   (b) the set $S = \{1/2^n : n \in \mathbb{N}\}$
   (c) the Cantor set Γ from Module 4
   (d) the complement of the Cantor set Γ

7. As in the exposition, let $h = h_4$. Graph h^2, h^3, and h^4. Make a conjecture about the graph of h^n and use your conjecture to prove that the periodic points of h are dense in $[0, 1]$.

8. Consider the dynamical system $([0, 1], f)$ where f is defined as

$$f(x) = \begin{cases} 2x & \text{if } x \leq \frac{1}{2} \\ 2x - 1 & \text{if } x > \frac{1}{2}. \end{cases}$$

   Form a conjecture as to whether or not $([0, 1], f)$ is chaotic. You need not rigorously prove your conjecture but give a clear explanation of the reasoning you used to reach it.

9. Prove Lemma 5.3.

10. Show that if (X, f) has a dense orbit then (X, f) is transitive.

Project

Determine whether the dynamical system $([0, 1], f)$, where $f(x) = 10x \pmod 1$, is chaotic. Prove your claim.

Module 6

Sharkovskii's Periodic Point Theorem

Exploration

Consider the dynamical systems $([0, 1], g_{3.5})$ and $([0, 1], g_4)$ where $g_{3.5}(x) = 3.5x(1-x)$ and $g_4(x) = 4x(1-x)$. Determine whether these systems have n-cycles for each positive integer n.

Exposition

Given a dynamical system (X, f), can we determine the values of $n \in \mathbb{N}$ for which it has an n-cycle? For any given value of n, we can find the fixed points of f^n by solving $f^n(x) = x$, but we will never be able to do this for every single $n \in \mathbb{N}$. Nevertheless, as we will see in this module, finding a complete answer is sometimes possible. We will discuss a clever technique used to show that sometimes the existence of an n-cycle for one value of n implies the existence of n-cycles for other values of n. We will illustrate this technique by investigating the n-cycles of the dynamical system $(\mathbb{R}, f)$ for the function $y = f(x)$ whose graph is shown in Figure 6.1.

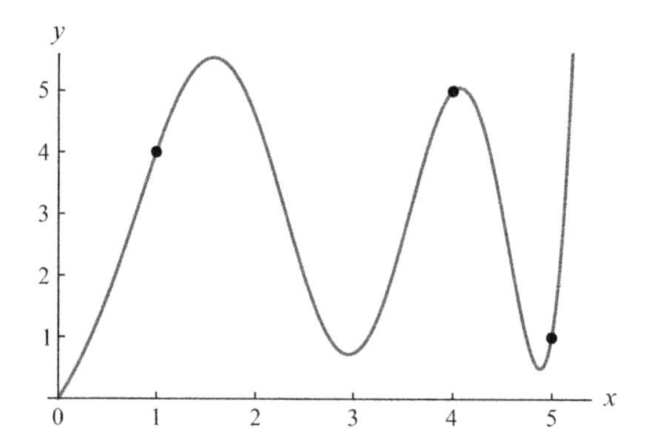

Figure 6.1. A graph of function f with its 3-cycle shown.

The dynamical system $(\mathbb{R}, f)$ has a 3-cycle consisting of $\{1, 4, 5\}$ with $f(1) = 4$, $f(4) = 5$, and $f(5) = 1$. Because f is a continuous function, the existence of this 3-cycle implies that the image of the interval $[1, 4]$ contains the interval $[4, 5]$, and the image of the interval $[4, 5]$ contains the interval $[1, 5]$. It is this property, together with the continuity of f, that will allow us to prove the existence of n-cycles for values of $n \in \mathbb{N}$ other than three. We will start by showing that $(\mathbb{R}, f)$ must have a 2-cycle.

Of course, the easiest way to show that there is a 2-cycle is to look at the graph of f^2. However, we are working towards a more general argument that depends on the existence of the 3-cycle. As we said above, $f([1, 4]) \supseteq [4, 5]$ and so we can choose a subinterval A of $[1, 4]$ whose image is exactly $[4, 5]$. Figure 6.2 shows one such subinterval.

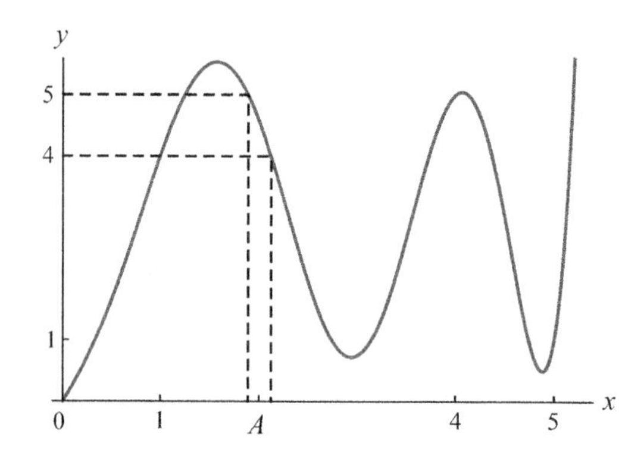

Figure 6.2. Subinterval A of $[1, 4]$ with $f(A) = [4, 5]$.

As we also said, we know that $f([4,5]) \supseteq [1,5]$ and thus $f^2(A) = f([4,5]) \supseteq [1,5]$. So f^2 stretches the subinterval A out to cover all of $[1,5]$. In particular, f^2 stretches A out over itself, so $f^2(A) \supseteq A$. This allows us to make use of the following property about continuous functions, which you will prove in Exercise 6.

Proposition 6.1. *Let h be a continuous function and I be a closed, bounded interval. If $h(I) \supseteq I$ then there exists a point $p \in I$ with $h(p) = p$.*

Using this proposition, we know there exists $p \in A$ with $f^2(p) = p$. Provided we can verify that p is not a fixed point for f, $\{p, f(p)\}$ will be the desired 2-cycle. To see that p is not a fixed point for f, we observe that $p \in A \subseteq [1,4]$ while $f(p) \in f(A) = [4,5]$. This means either $f(p) \neq p$ or $p = 4$. But $p \neq 4$ because the point 4 has minimal period three.

In this argument we did not find the 2-cycle directly. Nevertheless, we have demonstrated that at least one 2-cycle $\{p, f(p)\}$ with $p \in A$ must exist.

Can we extend this argument to show that the existence of the 3-cycle implies the existence of a 4-cycle? The most straightforward extension of the argument would be to start as before with a subinterval A of $[1,4]$ whose image is exactly $[4,5]$. As before, because $f(A) = [4,5]$, $f^2(A) = f([4,5])$ will include all of $[1,5]$. It then follows that $f^3(A)$ and $f^4(A)$ will also include all of $[1,5]$ and thus $A \subseteq f^4(A)$. We can then use Proposition 6.1 to deduce that there is a point $p \in A$ with $f^4(p) = p$. We need to show p has minimal period four before concluding that $\{p, f(p), f^2(p), f^3(p)\}$ is a 4-cycle. Arguing as before, p is not a fixed point. However, in showing p does not have period two, our straightforward extension of the earlier argument falls apart; since $f^2(A)$ contains A, we cannot rule out the possibility that p is also a period two point.

We will need to be a bit more careful in choosing our subinterval A if we want to conclude we have a 4-cycle. We would like to find a subinterval A of $[1,4]$ whose images under f, f^2, and f^3 all lie in $[4,5]$. This will ensure that no point in A (other than possibly the point 4) can have a period less than four. Then we will need the image of A under f^4 to contain all of $[1,5]$ and hence contain A. This will ensure the existence of a point $p \in A$ with $f^4(p) = p$. We know that $p \neq 4$ because the point 4 has minimal period three. Thus p has minimal period four.

To find a subinterval A as described above with $f(A)$, $f^2(A)$, and $f^3(A)$ all in $[4,5]$ and $f^4(A) \supseteq [1,5]$, it is easiest to work backwards. We start with $f^3(A)$. We want $f^3(A)$ to be a subinterval of $[4,5]$ with $f\left(f^3(A)\right) \supseteq [1,5]$. We can use $I_3 = [4,5]$ as a candidate for $f^3(A)$: clearly $I_3 = [4,5] \subseteq [4,5]$ and $f(I_3) \supseteq [1,5]$ as required.

If $I_3 = [4,5]$ is our candidate for $f^3(A)$, then I_2, our candidate for $f^2(A)$, must be a subinterval of $[4,5]$ that maps onto $I_3 = [4,5]$. The 3-cycle and continuity tell us that the image of $[4,5]$ includes all of $[1,5]$, so we know that such a subinterval will exist. Figure 6.3 illustrates a choice for I_2.

We work backwards again to find a candidate for $f(A)$. If I_2 is our candidate for $f^2(A)$, then I_1, our candidate for $f(A)$, must be a subinterval of $[4,5]$ that maps onto I_2. Figure 6.3 illustrates a choice for I_1.

We now have intervals I_1, I_2, I_3, all contained in $[4,5]$, with $f(I_1) = I_2$, $f(I_2) = I_3 = [4,5]$, and $f(I_3) = f([4,5]) \supseteq [1,5]$. If we can find an interval A in $[1,4]$ whose image is I_1, then our requirements for the images of A will be satisfied. But I_1 is in $[4,5]$ and the 3-cycle and continuity tell us that f maps $[1,4]$ onto $[4,5]$, so it will be possible to find an interval A in $[1,4]$ whose image is I_1. Figure 6.4 shows that there are two possible choices for A in our example. We have arbitrarily chosen one of them to call A.

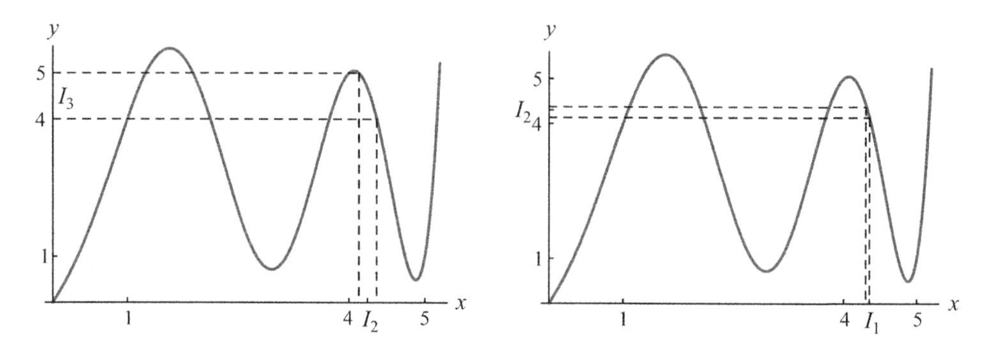

Figure 6.3. Finding candidates I_2 (left) and I_1 (right) for $f^2(A)$ and $f(A)$.

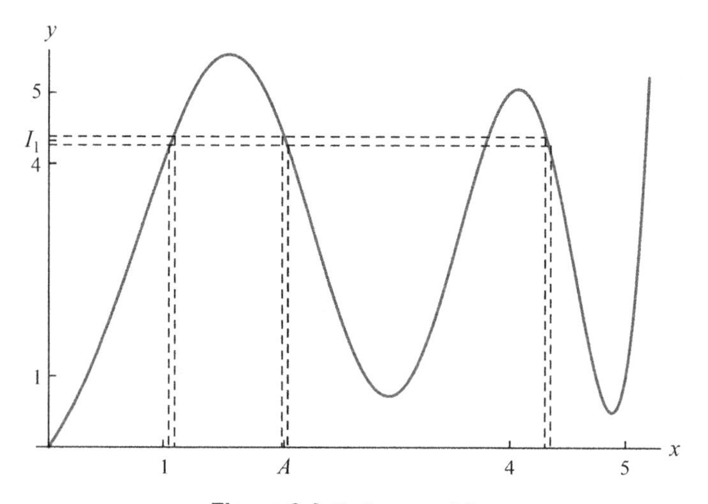

Figure 6.4. Preimages of I_1.

In summary, we have shown the existence of an interval $A \subseteq [1, 4]$ for which $f^4(A) \supseteq A$. Proposition 6.1 tells us that f^4 has a fixed point $p \in A$. We chose A so that $f(A)$, $f^2(A)$, and $f^3(A)$ are all contained in $[4, 5]$, and since p is not the number 4, p must have minimal period of four.

This argument, which used continuity and the existence of the 3-cycle to prove the existence of a 4-cycle, can be used to prove the existence of an n-cycle for any $n \in \mathbb{N}$. Given n, we will work backwards and find subintervals $I_j \subseteq [4, 5]$, $1 \leq j \leq n - 1$, with $f(I_j) = I_{j+1}$ for $1 \leq j \leq n - 2$ and $I_{n-1} = [4, 5]$. We will then choose interval $A \subseteq [1, 4]$ with $f(A) = I_1$, which we can do since continuity and the existence of the 3-cycle tell us f maps $[1, 4]$ onto $[4, 5]$. The subinterval $A \subseteq [1, 4]$ will then have the property that $f^n(A) \supseteq A$, and Proposition 6.1 tells us that A contains a point p with $f^n(p) = p$. However, with the possible exception of the point 4, there can be no periodic points in A of period less than n since $f^j(A) = I_j$ is contained in $[4, 5]$ for $1 \leq j \leq n - 1$ and A is contained in $[1, 4]$. Since we know the point 4 is part of the 3-cycle, as long as n is not a multiple of three, $p \neq 4$. Thus the orbit of p must be an n-cycle.

Now suppose n is a multiple of three. In this case $f^n(4) = 4$ so we need a different argument to exclude the possibility that $p = 4$. We will do this by considering the intervals I_{n-1}, I_{n-2}, and I_{n-3} more carefully. We choose the subinterval $I_{n-1} = [4, 5]$

as before. Because $f(5) = 1$, the subinterval I_{n-2} need not contain 5. So $I_{n-2} \subseteq [4,5)$. Then, because $f(4) = 5$, the subinterval I_{n-3} need not contain 4. So $I_{n-3} \subseteq (4,5)$. Each subsequent I_j, in particular I_1, is therefore also contained in $(4,5)$. The interval A, a subset of $[1,4]$ that maps onto I_1, will then be a proper subset of $(1,4)$. Thus $p \neq 4$ and p has minimum period n.

The discussion thus far of our particular example gives an idea of the flavor of the proof of the following theorem which is due to Tien-Yien Li and James Yorke [1] [LY].

Theorem 6.2. *Let $I \subseteq \mathbb{R}$ be an interval and let $f : I \to I$ be a continuous function. If (I, f) has a 3-cycle, then it has an n-cycle for all $n \in \mathbb{N}$.*

Thanks to Theorem 6.2, the question posed at the very beginning of the exposition can be answered when a dynamical system has a 3-cycle. Can we answer the question for dynamical systems with no 3-cycle? A remarkable theorem due to Oleksandr Sharkovskii[2] will help us to do so. Before stating the theorem, we give a definition that we will need.

Definition 6.3. Sharkovskii's ordering *on the natural numbers $\mathbb{N}$ is*

$$3 \rhd 5 \rhd 7 \rhd 9 \rhd \cdots \rhd 2\cdot 3 \rhd 2\cdot 5 \rhd 2\cdot 7 \rhd 2\cdot 9 \rhd \cdots \rhd$$
$$2^2\cdot 3 \rhd 2^2\cdot 5 \rhd 2^2\cdot 7 \rhd 2^2\cdot 9 \rhd \cdots \rhd 2^4 \rhd 2^3 \rhd 2^2 \rhd 2 \rhd 1.$$

We say that m is less than n in Sharkovskii's ordering if $n \rhd \cdots \rhd m$.

Sharkovskii's ordering is unusual, and in Exercise 1 you will study a few examples in order to understand the pattern. We are now ready to state Sharkovskii's theorem:

Theorem 6.4. (Sharkovskii's theorem) *Let $I \subseteq \mathbb{R}$ be an interval and let $f : I \to I$ be a continuous function. If the dynamical system (I, f) has an n-cycle, then it has an m-cycle for any m less than n in Sharkovskii's ordering.*

We will not prove Theorems 6.2 and 6.4, but interested readers can find more information in the original papers of Li and Yorke [LY] and Sharkovskii [S], or in a later work by Burns and Hasselblatt [BH].

Given $n \in \mathbb{N}$, Sharkovskii's theorem tells us what we may conclude from the existence of an n-cycle in a dynamical system, but it does not tell us if n-cycles exist. That is, given $n \in \mathbb{N}$, does there always exist a dynamical system with an n-cycle? Does there always exist a dynamical system for which the n-cycle is the largest cycle in terms of Sharkovskii's ordering? These questions will be explored in the project.

Exercises

1. Test your understanding of Sharkovskii's ordering:

   (a) List the numbers 1 through 16 from smallest to largest using Sharkovskii's ordering.

[1] James Yorke (b. 1941) is a Distinguished University Professor Emeritus with the Institute for Physical Science and Technology at the University of Maryland. He received the Japan Prize in Science and Technology in 2003 for his work on chaotic systems. Together with his student Tien-Yien Li (b. 1945), he coined the mathematical term chaos.

[2] Oleksandr Sharkovskii (b. 1936) was born in Kiev, Ukraine, where he won the Kiev Mathematical Olympiad as an 8th grader. He is currently a member of the faculty at the Institute of Mathematics of the Academy of Sciences in Kiev. He proved Sharkovskii's Theorem in 1964.

(b) If a dynamical system (X, f) has n-cycles for only finitely many $n \in \mathbb{N}$, what can you say about these n?

2. Use technology and Sharkovskii's theorem to determine for which values of $n \in \mathbb{N}$ the dynamical systems $([0,1], g_{3.5})$ and $([0,1], g_4)$ from the exploration have n-cycles.

3. A student makes the following statement:

   If a function f has a fixed point, then it has a periodic point of all periods. Therefore, there is a problem with Sharkovskii's ordering, because this implies that $1 \rhd q$ for any other q.

   What, if anything, is incorrect about this statement?

4. Show that Sharkovskii's theorem is not true for continuous maps from $\mathbb{R}^2$ to $\mathbb{R}^2$.

5. Does Sharkovskii's theorem hold if the function f is not required to be continuous? If so, explain why. If not, provide a counterexample.

6. (a) Draw a picture that illustrates why Proposition 6.1 is true.

   (b) Use the intermediate value theorem to prove Proposition 6.1.

7. The dynamical system $([0,1], g_4)$ has (approximately) the 3-cycle $\{0.117, 0.413, 0.970\}$.

   (a) Repeat the argument in the reading to show it has a 2-cycle. Give the intervals A and $f(A)$.

   (b) Repeat the argument in the reading to show it has a 4-cycle. Give the intervals A, $f(A)$, $f^2(A)$, and $f^3(A)$.

8. In this problem you will generalize the argument given in the reading. Let $h : \mathbb{R} \to \mathbb{R}$ be a continuous function, and suppose that the dynamical system $(\mathbb{R}, h)$ has a 3-cycle $\{a, b, c\}$ where $a < b < c$ and $h(a) = b$, $h(b) = c$, and $h(c) = a$.

   (a) Prove that $(\mathbb{R}, h)$ has a 2-cycle.

   (b) Prove that $(\mathbb{R}, h)$ has a 4-cycle.

   (c) Prove that $(\mathbb{R}, h)$ has an n-cycle for all $n \geq 1$.

Project

Order the natural numbers using Sharkovskii's ordering. Then Sharkovskii's theorem guarantees that, for each $n \in \mathbb{N}$, if (I, f) is a dynamical system with an n-cycle then (I, f) has an m-cycle for all smaller m. The theorem does not guarantee that for any particular value of n there is a dynamical system with an n-cycle and no larger cycles.

To prove that given $n \in \mathbb{N}$, there exists a function with an n-cycle and no larger cycles, we must give an example, or a procedure for generating an example, of such a function for each value of n.

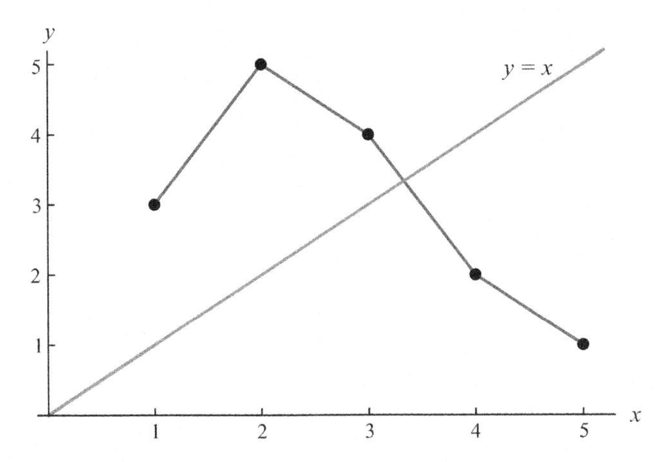

Figure 6.5. Graph of the function f.

1. Figure 6.5 is the graph of a function f defined on the interval $[1, 5]$ with the property that $f(1) = 3$, $f(3) = 4$, $f(4) = 2$, $f(2) = 5$, and $f(5) = 1$. Show that f does not have a 3-cycle.

2. Find a function with a 7-cycle but no 3-cycle or 5-cycle.

3. For any $n \in \mathbb{N}$, can the methods you used in constructing your example be used to construct a function with an n-cycle but no larger cycles?

Module 7

Bifurcations

Exploration

Consider the family of dynamical systems $(\mathbb{R}, g_c)$ where $g_c(x) = c\,x(1-x)$. Investigate how changing c changes the long term behavior of the initial condition $x_0 = 0.2$. At what values of c do the dynamics of the system experience a significant change?

Exposition

In Module 1 we modeled bacteria growth in a petri dish. The variable x took on values between zero and one and represented the proportion of the dish covered by bacteria at some point in time, and $g_2(x) = 2x(1-x)$ gave us the proportion of the dish covered one hour later. The term $(1-x)$ was included in the model to take into account the fact that, as the dish fills, crowding results in slower growth. The term $2x$ was included because in the absence of crowding, the amount of bacteria seemed to double every hour. Using this model, we predicted that any nonzero initial amount of bacteria would, with time, cover about half of the dish.

Would our prediction have been any different if instead of doubling every hour, the proportion of the dish covered by bacteria grew by two and a half times? Suppose it tripled every hour? After all, g_2 is just one of the functions in the family $g_c(x) = cx(1-x)$ which was studied in the exploration. In this context, $c = 2$ represents the growth rate of the bacteria and we can ask what would happen if our model used $c = 2.5$ or $c = 3$ instead of $c = 2$. Is there an optimal growth rate that will ensure that whatever the initial condition, our dish becomes as full as possible? Answering this last question entails studying the entire family g_c. Families of functions defined using a parameter arise often in applications. A second example is given in Exercise 5.

In this exposition we will consider how changing the parameter c affects the periodic point structure of functions in the family given by

$$h_c(x) = x^2 + x + c.$$

For any value of c, $y = h_c(x)$ is a parabola, opening up, with y-intercept c. Figure 7.1 shows the graphs of h_{-2}, h_0, and h_2 together with the line $y = x$; it appears from this illustration that h_{-2} has two fixed points, h_0 is tangent to the line $y = x$ and thus has a single fixed point, and h_2 has no fixed points at all.

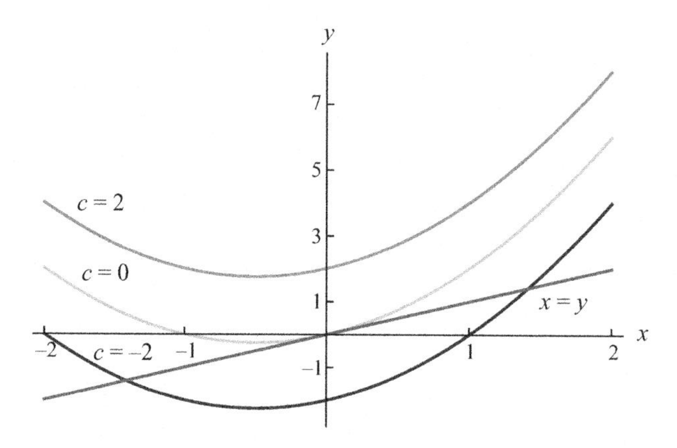

Figure 7.1. Graphs of h_{-2}, h_0, and h_2.

Figure 7.2 illustrates how as the parameter c increases from $c = -2$ to $c = 0$, the two fixed points for h_c seem to get closer and closer together, eventually coalescing at zero for $c = 0$. We can verify that this is the case by finding solutions to the equation

$$x^2 + x + c = x.$$

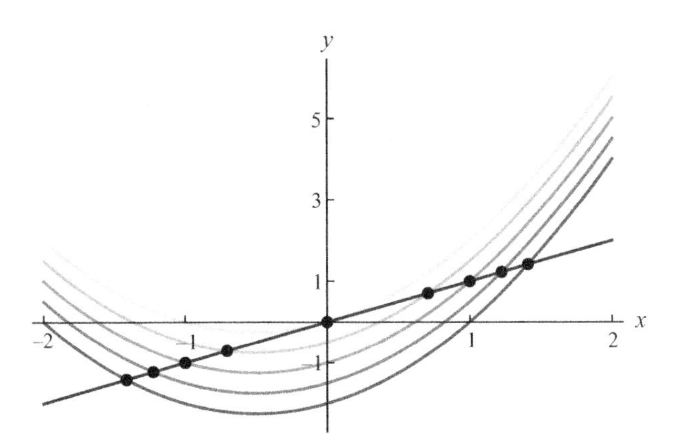

Figure 7.2. Graphs of h_c for c ranging from $c = -2$ to $c = 0$.

As suspected, this equation has two solutions $x = \pm\sqrt{-c}$ for $c < 0$, a single solution $x = 0$ for $c = 0$, and no solutions for $c > 0$.

We can investigate the nature of the fixed points of h_c for each value of c by considering the derivative of h_c at $\pm\sqrt{-c}$. In Exercise 4 you will verify that for $c < 0$, $p = \sqrt{-c}$ is always repelling while $p = -\sqrt{-c}$ is attracting for $-1 < c < 0$ and repelling otherwise.

At the value $c = 0$, we cannot use the derivative to determine the nature of the lone fixed point, $p = 0$, because

$$|h_0'(0)| = |2(0) + 1| = 1.$$

By examining iterates of initial conditions near $p = 0$, it appears that $p = 0$ is neither attracting nor repelling: initial conditions between -1 and 0 approach $p = 0$ and initial conditions greater than zero increase to infinity under iterations of h_0 (Module 1, Exercise 10).

A sudden change in the dynamics of a family of functions is called a **bifurcation** and the value of the parameter at which the change occurs is called a **bifurcation value**. In the example, a bifurcation occurs at $c = 0$ because the family goes from having two fixed points (one attracting and one repelling) for values of c that are less than zero to having no fixed points for values of c that are greater than zero. This type of bifurcation, where a pair of fixed points (one attracting and one repelling) appear as the parameter decreases (or increases), is called a **tangent bifurcation**.

By graphing select members of the family $\{h_c\}$ with the line $y = x$, Figures 7.1 and 7.2 illustrate the tangent bifurcation at $c = 0$. Two other types of graphs are also useful in illustrating how varying the parameter c in a family of functions can change important properties of the dynamical system. The first type of graph is called a **bifurcation diagram**, and it illustrates how the fixed points vary with c. To create a bifurcation diagram, we plot values of c on the horizontal axis and the corresponding fixed points on the vertical axis. The bifurcation diagram for the family of functions $h_c(x) = x^2 + x + c$ is shown in Figure 7.3.

The second type of useful graph is called an **orbit diagram**, and it illustrates how the long term behavior of a particular initial condition varies with c. In an orbit diagram, we again plot values for c on the horizontal axis. On the vertical axis, we plot what appears to be the eventual behavior of a fixed initial condition. In other words, we plot the iterates that occur after many iterations when we think the long term behavior of the initial condition has

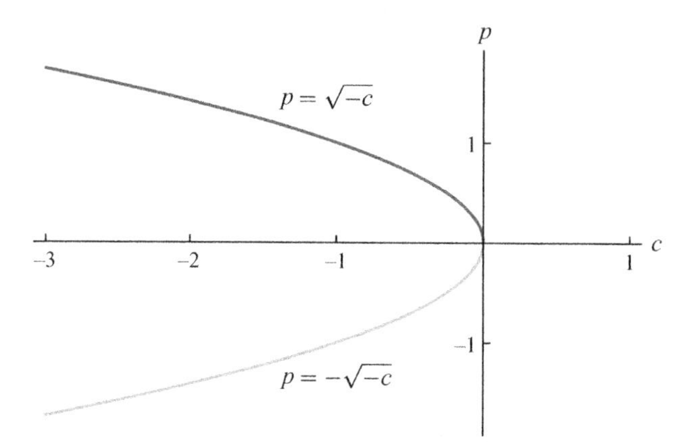

Figure 7.3. Bifurcation diagram for $h_c(x) = x^2 + x + c$.

stabilized. We will construct the orbit diagram, for the family h_c using the initial condition $x_0 = 0.2$. In Exercise 2 you will construct orbit diagrams for this same family using other choices for the initial condition.

For the family h_c, because $x^2 + x \geq x$ for all values of x, it is clear that if $c > 0$, iterates of any initial condition will go off to infinity. So in creating an orbit diagram for the family h_c, we will only concern ourselves with $c \leq 0$. Table 7.1 shows iterates h_c^{100} through h_c^{104} for initial condition $x_0 = 0.2$ for a few negative values of c.

n\c	−0.25	−0.75	−1.25	−1.6
100	−0.5	−0.87	−1.5	−0.04
101	−0.5	−0.87	−0.5	−1.64
102	−0.5	−0.87	−1.5	−0.56
103	−0.5	−0.87	−0.5	−1.85
104	−0.5	−0.87	−1.5	−0.04

Table 7.1. Values of h_c^n for initial condition $x_0 = 0.2$.

From the table, it appears that for $c = -0.25$ and $c = -0.75$, the iterates of $x_0 = 0.2$ approach −0.5 and −0.87 respectively; for $c = -1.25$, the iterates oscillate between −1.5 and −0.5; and for $c = -1.6$, the iterates oscillate between four values: −0.04, −1.64, −0.56, and −1.85. This observation is illustrated in the partial orbit diagram as shown in Figure 7.4.

To complete the orbit diagram, we continue in this way for many other values of c resulting in the diagram shown in Figure 7.5.

From the orbit diagram in Figure 7.5, it appears that the long term behavior of the iterates of $x_0 = 0.2$ changes radically at $c = -1$ and at $c = -1.5$. Consider first the change in behavior at $c = -1$. For $-1 \leq c < 0$, the iterates of $x_0 = 0.2$ appear to approach an attracting fixed point, but for $c < -1$, the iterates of $x_0 = 0.2$ appear to approach an attracting 2-cycle. We have already discussed the fact that the fixed point $p = -\sqrt{-c}$ is attracting for $-1 < c < 0$ (Exercise 4), so the first part of this behavior is not unexpected.

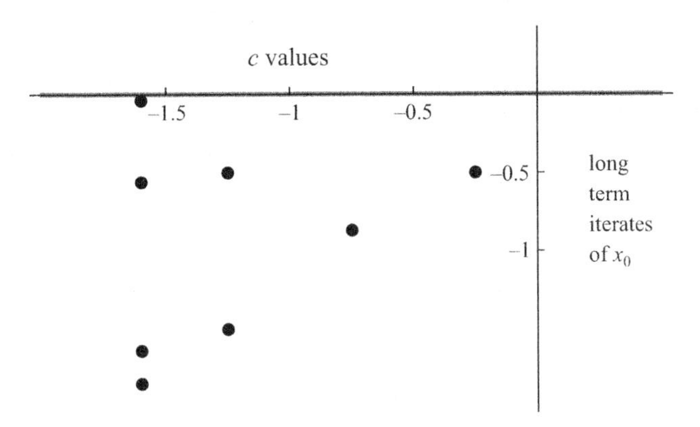

Figure 7.4. Orbit diagram for $x_0 = 0.2$ and $c = -0.25, -0.75, -1.25$, and -1.6.

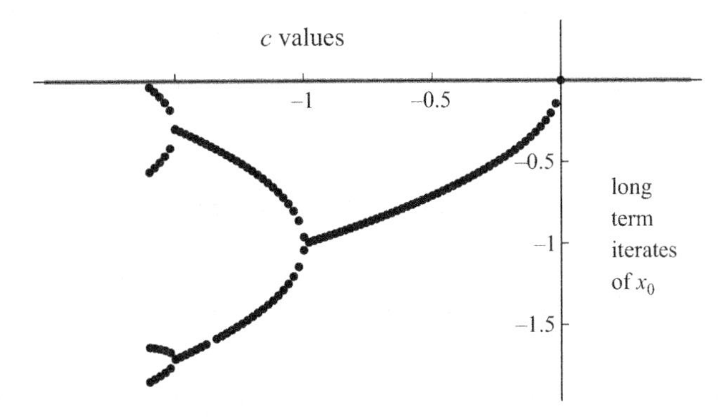

Figure 7.5. Orbit diagram for $x_0 = 0.2$ and $-1.6 \leq c \leq 0$.

We turn our attention to the 2-cycles of h_c. The equation $h_c(h_c(x)) = x$ has solutions

$$x = -1 \pm \sqrt{-1 - c},$$

so h_c has a 2-cycle $\{p, q\} = \{-1 + \sqrt{-1 - c},\ -1 - \sqrt{-1 - c}\}$ whenever $c < -1$. This 2-cycle will be attracting whenever

$$\left| \frac{d}{dx}(h_c(h_c(p))) \right| < 1.$$

Solving this inequality tells us that we have an attracting 2-cycle for $-1.5 < c < -1$ and a repelling 2-cycle for $c < -1.5$ (Exercise 4). Thus, as we suspected, there is a second bifurcation at $c = -1.5$. For values of c less than $c = -1.5$, the iterates of $x_0 = 0.2$ seem to approach an attracting 4-cycle, while for values of c greater than $c = -1.5$ (but still less than $c = -1$), the iterates of $x_0 = 0.2$ approach an attracting 2-cycle. We can verify the nature of the 4-cycle for $c < -1.5$ by investigating the fixed points of h_c^4 just as we investigated the fixed points of h_c^2.

A value of the parameter c at which an n-cycle turns into a $2n$-cycle is called a **period doubling bifurcation**. The family of functions h_c undergoes a period doubling bifurcation at $c = -1$ and $c = -1.5$. In the project, you will find even more period doubling bifurcations for this family.

Mathematical models are being used with increasing frequency to inform decisions, and clearly if a mathematical model consists of a family of functions depending on a parameter, knowledge of the bifurcations that can occur at various parameter values is critical to using the model to predict long term behavior. In Exercise 5, you will see a simple example of how a model might be used and of why understanding its bifurcations is so important.

Exercises

1. The following short answer questions test your understanding of the orbit diagram for the family of functions h_c found in Figure 7.5.

   (a) Based on Figure 7.5, what would you expect if you were to iterate h_c for $x_0 = 0.2$ and $c = -0.5$? For $c = -1.25$? For $c = -1.6$?

   (b) Is the following statement true or false?

   Figure 7.5 shows us that for $c < -1$ there are no fixed points, only period two (or greater) cycles.

   (c) Is the following statement true or false?

   Figure 7.5 shows us that $h_c(-0.5)$ is approximately -0.75.

   (d) Is the following statement true or false?

   Figure 7.5 shows us that $h_c(-0.3)$ is approximately -1.7 for $c = -1.5$.

2. The initial condition $x_0 = 0.2$ used in creating the orbit diagram in the exposition was chosen randomly. Create a variety of orbit diagrams for the family h_c using different initial conditions.

3. Make an orbit diagram for the family of functions found in the exploration.

4. (a) For $c < 0$, the function h_c has fixed points at $p = \pm\sqrt{-c}$. Determine their nature as a function of c.

   (b) For $c < -1$, the function h_c has a 2-cycle

   $$\{p, q\} = \{-1 + \sqrt{-1-c},\ -1 - \sqrt{-1-c}\}.$$

   Determine its nature as a function of c.

5. Suppose that bacteria is grown in a petri dish, and the model describing its population growth is $g_2(x)$. Suppose also that a certain amount of the bacteria in the petri dish is harvested for medicinal purposes every unit of time. We might modify the model as:

   $$r_p(x) = 2x(1-x) - p.$$

   (a) For $p = 0.1$, find the long term behavior for all initial conditions. Interpret your answer in context.

   (b) Find a bifurcation value for p and interpret the importance of this bifurcation in context.

   (c) What would be your recommendation for harvesting bacteria?

6. We can think of the period doubling bifurcation in the family h_c at $c = -1$ as a tangent bifurcation in the family h_c^2. Plot h_c^2 for various values of c (as in Figure 7.2) in order to illustrate this tangent bifurcation. Explain the connection between the tangent bifurcation in the family h_c^2 and the period doubling bifurcation in the family h_c.

Project

In this project you will expand the orbit diagram for the family of functions h_c. We have found (Exercise 4) that there is a period doubling bifurcation at $c = -1$. Call this c_1. From the orbit diagram (Figure 7.5) we see there is another period doubling bifurcation at $c = -1.5$: call this c_2. Now find the period doubling bifurcation where the family goes from having an attracting 4-cycle to an attracting 8-cycle. Do this by iterating the initial condition $x_0 = 0.2$ for values of c at regularly spaced intervals and looking for changes in long term behavior. (Hint: Based on Figure 7.5, do you have an idea of how small the interval will need to be in order to detect the bifurcation? Should it be less or greater than 0.5?) Call this value c_3. Continue in this way to find other period doubling bifurcations. Illustrate your findings in an orbit diagram, changing the scale as needed.

Next, define the sequence $\{x_i\}$ by setting $x_i = c_{i+1} - c_i$. Discuss what you can say about $\lim_{i\to\infty} \dfrac{x_i}{x_{i+1}}$.

Finally, do the same for the family of functions defined in the exploration and discuss any relationships you find.

Module 8

Conjugacy of Dynamical Systems

Exploration

Compare the long term behavior of initial conditions in each of the pairs of dynamical systems given below. In what ways are they similar or different as dynamical systems?

1a. $(\mathbb{R}, f_1)$ where $f_1(x) = \frac{1}{2}x$.
1b. $(\mathbb{R}, f_2)$ where $f_2(x) = 2x$.

2a. $([0, 1], h)$ where $h(x) = \begin{cases} 2x & \text{if } x \le \frac{1}{2} \\ 2 - 2x & \text{if } x > \frac{1}{2}. \end{cases}$

2b. $([0, 1], g_4)$ where $g_4(x) = 4x(1 - x)$.

Exposition

Previous modules have focused on identifying different types of long term behavior in dynamical systems. For example, in Module 1 we studied the family of dynamical systems $(\mathbb{R}, g_c)$ where $g_c(x) = cx(1-x)$, and we saw that for some values of c the system had an attracting fixed point. For other values of c, the system had repelling fixed points and an attracting two-cycle, and for yet other values of c the system had no attracting cycles at all. So different values of c can yield fundamentally different dynamical systems despite the similarities in the defining functions. On the other hand, the exploration suggests that two dynamical systems can be similar even if their defining functions are different. When two systems have the same types of orbits, it is natural to say that from a dynamical systems point of view they are identical. In this module, we will make this idea mathematically precise. In later modules, we will occasionally exploit the tools we develop here; we will avoid having to work hard to understand a new dynamical system by recognizing that it is identical to one that we have already studied.

To give a rigorous definition of when two dynamical systems are identical, we first need a correspondence between the points in one phase space and the points in the other phase space. This correspondence should preserve long term behavior in the sense that a periodic point in the first phase space should correspond to a periodic point in the second phase space, a point with a bounded orbit in the first phase space should correspond to a point with a bounded orbit in the second phase space, an attracting fixed point in the first phase space should correspond to an attracting fixed point in the second phase space, and so on.

We begin by considering the type of function we would need in order to associate the orbit of a point from the first phase space to the orbit of a point in the second phase space.

Definition 8.1. *Let (X, f) and (Y, g) be dynamical systems. A function $\phi : X \to Y$* **commutes with f and g** *if $\phi(f(x)) = g(\phi(x))$ for every $x \in X$. This is illustrated in the diagram:*

$$\begin{array}{ccc} X & \xrightarrow{f} & X \\ \phi \downarrow & & \downarrow \phi \\ Y & \xrightarrow{g} & Y. \end{array}$$

In Exercise 1 you will show that such a function ϕ maps the orbit of a point x in X to the orbit of $\phi(x)$ in Y. In Exercise 2 you will prove the next result which states, in particular, that such a map ϕ must map periodic points to other periodic points.

Lemma 8.2. *Let (X, f) and (Y, g) be dynamical systems and let $\phi : X \to Y$ be a function that commutes with f and g. If $x \in X$ has period n, then $\phi(x) \in Y$ is a periodic point and its minimal period divides n.*

The converse of this lemma need not be true; there are functions ϕ satisfying the hypotheses of this lemma that map non-periodic points to periodic points. For example, consider $([0, 1], g_4)$ and (Y, f) where the phase space $Y = \{0\}$ consists of a single point and $f(0) = 0$. Define $\phi : [0, 1] \to Y$ by $\phi(x) = 0$ for all $x \in [0, 1]$. Namely, ϕ maps every point in $[0, 1]$ to 0. As shown below in a pointwise version of the commuting diagram from Definition 8.1, the function ϕ commutes with the maps g_4 and f, illustrating that it satisfies the hypotheses of Lemma 8.2:

$$\begin{array}{ccc} x & \overset{g_4}{\mapsto} & g_4(x) \\ \phi \downarrow & & \downarrow \phi \\ 0 & \overset{f}{\mapsto} & 0. \end{array}$$

However, ϕ maps every orbit in $([0,1], g_4)$ to the fixed point 0 despite the fact that $([0,1], g_4)$ has many different kinds of orbits, including n-cycles and infinite orbits. The problem is that although $\phi : [0,1] \to Y$ commutes with g_4 and f, and thus maps orbits to orbits, it is not one-to-one. When a commuting map ϕ is both one-to-one and onto, it is an invertible function, and ϕ^{-1} is also a commuting map (Exercise 3). By applying Lemma 8.2 to both ϕ and ϕ^{-1}, it follows that there is a one-to-one correspondence between periodic orbits of the same periods in the two systems. The next lemma, the proof of which is the subject of Exercise 4, is a formal statement of this fact.

Lemma 8.3. *Let (X, f) and (Y, g) be dynamical systems and let $\phi : X \to Y$ be a one-to-one and onto function that commutes with f and g. Then $x \in X$ has minimal period n if and only if $\phi(x) \in Y$ has minimal period n.*

We have now established that a one-to-one, onto, commuting function between two dynamical systems guarantees a one-to-one correspondence between orbits in which the periods of finite orbits are preserved. However, the exploration shows us that such a correspondence between orbits is not enough to guarantee similar long term behaviors. To see this, consider $\phi : \mathbb{R} \to \mathbb{R}$ given by

$$\phi(x) = \begin{cases} \frac{1}{x} & \text{if } x \neq 0 \\ 0 & \text{if } x = 0. \end{cases}$$

Note that ϕ is one-to-one and onto. It is easy to check that ϕ commutes with the functions f_1 and f_2 as defined in the exploration; if $x = 0$, then the claim is obvious. If $x \neq 0$, we have

$$\phi(f_1(x)) = \phi\left(\frac{1}{2}x\right) = \frac{2}{x} = 2\left(\frac{1}{x}\right) = f_2\left(\frac{1}{x}\right) = f_2(\phi(x)).$$

Both $(\mathbb{R}, f_1)$ and $(\mathbb{R}, f_2)$ have exactly one fixed point and no other cycles. Given that ϕ is a commuting, one-to-one, onto function we are guaranteed that it will map the fixed point to the fixed point and infinite orbits to infinite orbits. However, the fixed point in the first system is attracting while the fixed point in the second is repelling, and all infinite orbits in the first system approach the fixed point but are mapped by ϕ to infinite orbits that are unbounded. Thus, although a one-to-one, onto, commuting function preserves the structure of an n-cycle, it may not preserve its nature (attracting, repelling, neither).

In the example, the problem lies with the discontinuity of ϕ at the fixed point $x = 0$. When a function is discontinuous, the image of a converging sequence in the domain need not converge in the range. This is what happens in the example when the orbits that converge to the attracting fixed point in $(\mathbb{R}, f_1)$ have images that do not converge in $(\mathbb{R}, f_2)$. So continuity is a condition we will need to impose in order to characterize the type of functions that will create a correspondence between orbits and preserve the long term behaviors in two identical dynamical systems. This is the final condition that will be needed.

Definition 8.4. *A function $\phi : X \to Y$ is a* **homeomorphism** *if it is one-to-one, onto, continuous, and its inverse function ϕ^{-1} is continuous.*

We are now ready to give a rigorous mathematical definition of identical, or conjugate, dynamical systems:

Definition 8.5. *The dynamical systems (X, f) and (Y, g) are* **conjugate** *if there exists a homeomorphism $\phi : X \to Y$ that commutes with f and g. We call the function ϕ a* **conjugacy**.

As established in Exercise 1, a conjugacy creates a correspondence between the orbits of two systems. Lemma 8.3 and Theorem 8.6 (below) tell us that this correspondence preserves both the period and nature of n-cycles.

Theorem 8.6. *Let (X, f) and (Y, g) be dynamical systems and let $\phi : X \to Y$ be a conjugacy. If the orbit of $p \in X$ is an attracting (repelling, neither) n-cycle, then the orbit of $\phi(p) \in Y$ is an attracting (repelling, neither) n-cycle as well.*

We will not give a full proof of Theorem 8.6. Instead we will prove the special case that is given in the following proposition. Another special case is considered in Exercise 7. The full proof uses similar ideas.

Proposition 8.7. *Let (X, f) and (Y, g) be dynamical systems and let $\phi : X \to Y$ be a conjugacy. If $p \in X$ is an attracting fixed point of (X, f), then $\phi(p)$ is an attracting fixed point of (Y, g).*

Proof. Let $p \in X$ be an attracting fixed point. This means that there exists $\epsilon > 0$ such that for all $x \in X$ with $|p - x| < \epsilon$, we have $\lim_{n\to\infty} f^n(x) = p$. In order to show that $\phi(p)$ is also attracting, we need to show the same holds true for $\phi(p)$: that is, there exists another $\epsilon > 0$, call it $\tilde{\epsilon}$, such that for all $y \in Y$ with $|\phi(p) - y| < \tilde{\epsilon}$, $\lim_{n\to\infty} g^n(y) = \phi(p)$.

We will use the continuity of ϕ^{-1} to give us a value for $\tilde{\epsilon}$. Since ϵ is fixed, the continuity of ϕ^{-1} tells us that there exists $\tilde{\epsilon} > 0$ so that if $|\phi(p) - y| < \tilde{\epsilon}$, then $\left|p - \phi^{-1}(y)\right| < \epsilon$. We now check that this $\tilde{\epsilon}$ satisfies the definition of $\phi(p)$ being an attracting fixed point.

Let $y \in Y$ be such that $|\phi(p) - y| < \tilde{\epsilon}$. We then know that $\left|p - \phi^{-1}(y)\right| < \epsilon$. Since p is an attracting fixed point, $\lim_{n\to\infty} f^n\left(\phi^{-1}(y)\right) = p$. By the continuity of ϕ we have

$$\lim_{n\to\infty} \phi\left(f^n\left(\phi^{-1}(y)\right)\right) = \phi(p).$$

Since ϕ commutes with f and g, we can rewrite this as

$$\lim_{n\to\infty} f^n\,(y) = \phi(p).$$

This gives us the desired result. □

We close this module by proving that the two dynamical systems $([0, 1], h)$ and $([0, 1], g_4)$ defined in the exploration are conjugate. In order to do so, consider the function $\phi(x) = \sin^2(\frac{\pi}{2}x)$ and its inverse, whose graphs are shown in Figure 8.1. These graphs suggest that both ϕ and ϕ^{-1} are continuous, one-to-one, and onto functions from $[0, 1]$ to $[0, 1]$; Exercise 6b requires a rigorous proof of these assertions. Thus ϕ is a homeomorphism, and to show that it is a conjugacy, we need only show it commutes with h and g_4. That is, we need to show that $\phi(h(x)) = g_4(\phi(x))$ as illustrated in the following commuting diagram:

$$\begin{array}{ccc} [0,1] & \stackrel{h}{\to} & [0,1] \\ \phi\downarrow & & \downarrow\phi \\ [0,1] & \stackrel{g_4}{\to} & [0,1]. \end{array}$$

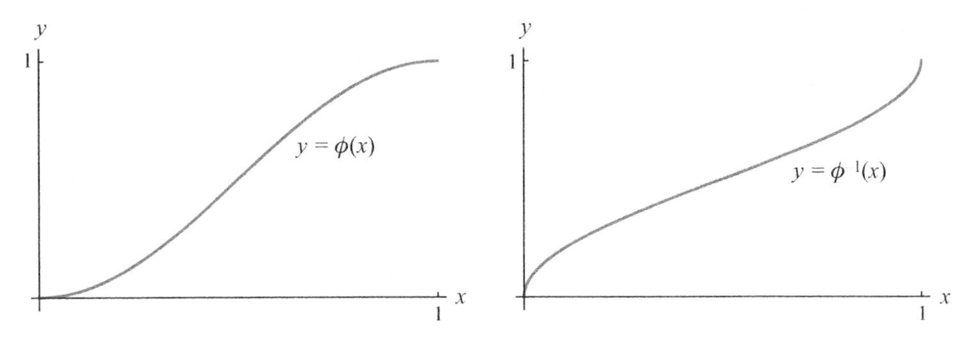

Figure 8.1. Graph of $\phi(x) = \sin^2(\frac{\pi}{2}x)$ and its inverse.

Let $x \in [0, 1]$ and apply ϕ and then g_4, as shown:

$$\begin{array}{ccc} & x & \\ \phi & \downarrow & \\ & \sin^2(\frac{\pi}{2}x) & \stackrel{g_4}{\longmapsto} \quad 4(\sin^2(\frac{\pi}{2}x))\,(1-\sin^2(\frac{\pi}{2}x)). \end{array}$$

In Exercise 6a, you will use the trigonometric identity $\sin^2\theta + \cos^2\theta = 1$ and the double angle formula $\sin(2\theta) = 2\sin\theta\,\cos\theta$ to rewrite $4(\sin^2(\frac{\pi}{2}x))\,(1-\sin^2(\frac{\pi}{2}x))$ as $\sin^2(\pi x)$. This simplifies the diagram to

$$\begin{array}{ccc} & x & \\ \phi & \downarrow & \\ & \sin^2(\frac{\pi}{2}x) & \stackrel{g_4}{\longmapsto} \quad \sin^2(\pi x). \end{array}$$

Now start with the same x and apply first h and then ϕ. Because of the definition of h, we need to consider two cases:

Case 1: $x \leq 1/2$

$$\begin{array}{ccc} x & \stackrel{h}{\longmapsto} & 2x \\ & & \downarrow \quad \phi \\ & & \sin^2(\pi x). \end{array}$$

Case 2: $x > 1/2$

$$\begin{array}{ccc} x & \stackrel{h}{\longmapsto} & 2-2x \\ & & \downarrow \quad \phi \\ & & \sin^2(\pi - \pi x). \end{array}$$

The sine function is periodic and so $\sin^2(\pi - \pi x) = \sin^2(\pi x)$, giving us the commutative diagram for all $x \in [0, 1]$:

$$\begin{array}{ccc} x & \stackrel{h}{\longmapsto} & 2x \text{ or } 2-2x \\ \phi \downarrow & & \downarrow \phi \\ \sin^2(\frac{\pi}{2}x) & \stackrel{g_4}{\longmapsto} & \sin^2(\pi x). \end{array}$$

It may seem like the function ϕ came out of nowhere. Indeed, determining the conjugacy function is the challenging part of showing that two dynamical systems are conjugate, but in the project you will see that the rewards of establishing a conjugacy are great. The full power of conjugacies will be exploited in Modules 11 and 14.

Exercises

1. Let (X, f) and (Y, g) be dynamical systems and suppose that $\phi : X \to Y$ commutes with f and g. Prove that ϕ maps the orbit of a point x in X to the orbit of $\phi(x)$ in Y.

2. Prove Lemma 8.2.

3. Let (X, f) and (Y, g) be two dynamical systems. Let $\phi : X \to Y$ be a one-to-one, onto map that commutes with f and g. Show that $\phi^{-1} : Y \to X$ commutes with g and f.

4. Prove Lemma 8.3.

5. Let $\phi : X \to Y$ be a conjugacy between (X, f) and (Y, g). Show that for each $n \in \mathbb{N}$ and $y \in Y$, $\phi(f^n(\phi^{-1}(y))) = g^n(y)$.

6. This exercise involves proofs of statements made in the exposition.

   (a) Use various trigonometric identities to show that

   $$4(\sin^2(\frac{\pi}{2}x))\,(1 - \sin^2(\frac{\pi}{2}x)) = \sin^2(\pi x).$$

   (b) Prove that $\phi(x) = \sin^2(\frac{\pi}{2}x)$ and $\phi^{-1}(x)$ are continuous, one-to-one, onto functions from $[0, 1]$ to $[0, 1]$.

7. Let (X, f) and (Y, g) be dynamical systems, and let $\phi : X \to Y$ be a conjugacy. Prove that $p \in X$ is a repelling fixed point if and only if $\phi(p) \in Y$ is a repelling fixed point.

8. Let $f(x) = 8x$ and $g(x) = 2x$. Show that $\phi(x) = x^{\frac{1}{3}}$ is a conjugacy between $(\mathbb{R}, f)$ and $(\mathbb{R}, g)$.

9. Let f and g be given by

   $$f(x) = \begin{cases} 3x & \text{if } x \le \frac{1}{2} \\ 3 - 3x & \text{if } x > \frac{1}{2}, \end{cases} \qquad g(x) = \begin{cases} 3x - 2 & \text{if } x \le 2 \\ -3x + 10 & \text{if } x > 2. \end{cases}$$

   Show that $\phi : \mathbb{R} \to \mathbb{R}$ defined by $\phi(x) = 2x + 1$ is a conjugacy between $(\mathbb{R}, f)$ and $(\mathbb{R}, g)$.

10. Let ϕ be a conjugacy between (X, f) and (Y, g). Show that if f is one-to-one then g is one-to-one.

11. A function $f : X \to Y$ is **uniformly continuous** if for every $\epsilon > 0$ there exists $\delta > 0$ such that if $x, y \in X$ satisfies $|x - y| < \delta$ then $|f(x) - f(y)| < \epsilon$. Uniform continuity is stronger than the usual notion of continuity since the value of δ is independent of the values of x and y.

    (a) Show that $f(x) = x^2$ is uniformly continuous on $[0, 1]$.

    (b) Show that $\phi(x) = 1/x$ is not uniformly continuous on $(0, 1)$, but that it is uniformly continuous on $[1/2, 1]$.

    This definition will be useful in the project.

Project

Let X and Y be closed and bounded intervals, and assume that (X, f) and (Y, g) are conjugate systems. Show that if (X, f) is chaotic then so is (Y, g).

You may find it useful to use the following result:

Theorem 8.8. *Let X and Y be closed and bounded intervals. If $h : X \to Y$ is a continuous function then h is uniformly continuous.*

Module 9

Two-Dimensional Discrete Dynamical Systems

Exploration

Find all possible long term behaviors of initial conditions for the dynamical system $(\mathbb{R}^2, f)$ where

$$f(x, y) = \left(\frac{5}{4}x - \frac{3}{4}y, -\frac{3}{4}x + \frac{5}{4}y\right).$$

Exposition

We have thus far studied dynamical systems (X, f) where the phase space X is a subset of $\mathbb{R}$. We have viewed an initial condition $x_0 \in X$ as representing a quantity, such as population size, that varies with time. However, many physical systems require a more complicated model. For example, suppose that we want to model the movement of an object on earth over time. We could represent the position of the object by a two-tuple of real numbers describing the latitude and longitude of the position. The model of the movement of the object is then a function $f : \mathbb{R}^2 \to \mathbb{R}^2$. In this case, if the object is initially at location $(x_0, y_0) \in \mathbb{R}^2$, $f(x_0, y_0)$ gives the location after one time unit, $f^2(x_0, y_0)$ gives the location after two time units, and so on.

We can easily extend the definition of a fixed point to dynamical systems of the form $(\mathbb{R}^2, f)$.

Definition 9.1. *A point (p, q) in the domain of a function $f : \mathbb{R}^2 \to \mathbb{R}^2$ is called a* **fixed point** *of f if $f(p, q) = (p, q)$.*

Moreover, using the familiar notion of the distance between two points (x, y), (u, v) in $\mathbb{R}^2$ given by

$$d\left((x, y), (u, v)\right) = \sqrt{(x-u)^2 + (y-v)^2},$$

we can define what it means for a fixed point to be attracting and repelling just as we did for systems on $\mathbb{R}$.

Definition 9.2. *Let (p, q) be a fixed point of the dynamical system $(\mathbb{R}^2, f)$. We say (p, q) is an* **attracting fixed point** *if there exists $\epsilon > 0$ such that every $(x, y) \in \mathbb{R}^2$ with $d((p, q), (x, y)) < \epsilon$ satisfies $\lim_{n\to\infty} f^n(x, y) = (p, q)$.*

As before, if (p, q) is an attracting fixed point, the set of initial conditions (x_0, y_0) with the property that $\lim_{n\to\infty} f^n(x_0, y_0) = (p, q)$ is called the **basin of attraction** of (p, q).

Definition 9.3. *Let (p, q) be a fixed point of the dynamical system $(\mathbb{R}^2, f)$. We say (p, q) is a* **repelling fixed point** *if there exists $\epsilon > 0$ such that for every $(x, y) \neq (p, q) \in \mathbb{R}^2$ with $d((p, q), (x, y)) < \epsilon$, there exists a positive integer k such that $d(f^k(x, y), (p, q)) \geq \epsilon$.*

We illustrate these ideas with two examples. Let $f_1(x, y) = (x/2, y/3)$ and $f_2(x, y) = (2x, 3y)$. We see that $(0, 0)$ is a fixed point for both $(\mathbb{R}^2, f_1)$ and $(\mathbb{R}^2, f_2)$. Figure 9.1 suggests that $(0, 0)$ is an attracting fixed point for the first system and a repelling fixed point for the second system. In fact, we see that for any initial condition (x_0, y_0),

$$\lim_{k\to\infty} f_1^k(x_0, y_0) = \lim_{k\to\infty} \left(\frac{1}{2^k}x_0, \frac{1}{3^k}y_0\right) = (0, 0).$$

So any choice of $\epsilon > 0$ will work in Definition 9.2, and the basin of attraction of $(0, 0)$ is all of $\mathbb{R}^2$. On the other hand, since $f_2^k(x_0, y_0) = (2^k x_0, 3^k y_0)$, we see that for any initial condition $(x_0, y_0) \neq (0, 0)$ the distance between $f_2^k(x_0, y_0)$ and $(0, 0)$ grows as k increases to infinity. Thus any choice of $\epsilon > 0$ will work in Definition 9.3 and $(0, 0)$ is a repelling fixed point for f_2. More precisely, if we set d_k to be the distance between $f_2^k(x_0, y_0)$ and $(0, 0)$ we see that $\lim_{k\to\infty} d_k = \infty$. In this case we say the sequence of points $\{f_2^k(x_0, y_0)\}$ in $\mathbb{R}^2$ **grows without bound**. For a sequence $\{(x_k, y_k)\}$ to grow without bound, it is not necessary for both coordinates to tend to infinity.

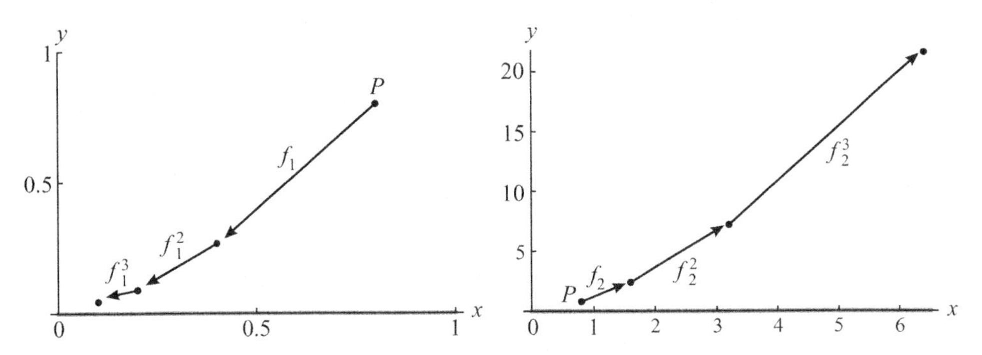

Figure 9.1. Examples of iterates of $P = (x_0, y_0)$ for f_1 and f_2.

We have seen that some fixed points are neither attracting nor repelling, and this can also occur for functions from $\mathbb{R}^2$ to $\mathbb{R}^2$. For example, consider the function $f_3(x, y) = (x/2, 3y)$. In this case, $f_3^k(x, y) = \left(x/2^k, 3^k y\right)$ and initial conditions on the x-axis approach the fixed point $(0, 0)$ under iteration, while all other initial conditions grow without bound. Thus neither Definition 9.2 nor Definition 9.3 apply, and the fixed point $(0, 0)$ is neither attracting nor repelling.

All the functions we have discussed thus far are examples of linear maps. **Linear maps** $f : \mathbb{R}^2 \to \mathbb{R}^2$ are those of the form

$$f(x, y) = (ax + by, cx + dy)$$

where a, b, c, and d are constants. Linear maps can be understood completely if we use the vector space structure of $\mathbb{R}^2$ and represent them as matrices. In particular, given any linear map $f : \mathbb{R}^2 \to \mathbb{R}^2$ there is a 2×2 matrix $A = \begin{bmatrix} a & b \\ c & d \end{bmatrix}$ so that $f(x_0, y_0) = (x_1, y_1)$ if and only if $A \begin{bmatrix} x_0 \\ y_0 \end{bmatrix} = \begin{bmatrix} x_1 \\ y_1 \end{bmatrix}$. For instance, the function f_3 can be represented by the matrix $\begin{bmatrix} \frac{1}{2} & 0 \\ 0 & 3 \end{bmatrix}$. Note that $(0, 0)$ will be a fixed point for any linear map. Conditions on the matrix A will determine whether it is the only fixed point or if there are others (Exercise 4).

For the dynamical systems $(\mathbb{R}^2, f_1)$, $(\mathbb{R}^2, f_2)$, and $(\mathbb{R}^2, f_3)$, any initial condition that starts on an axis will remain on it after iteration. The axes are examples of sets that are **invariant** under the map, and the invariance is a consequence of properties of the matrices associated with f_1, f_2, and f_3. In particular, the associated matrix for f_3 satisfies

$$\begin{bmatrix} \frac{1}{2} & 0 \\ 0 & 3 \end{bmatrix} \begin{bmatrix} 1 \\ 0 \end{bmatrix} = \begin{bmatrix} \frac{1}{2} \\ 0 \end{bmatrix} = \frac{1}{2} \begin{bmatrix} 1 \\ 0 \end{bmatrix}$$

and

$$\begin{bmatrix} \frac{1}{2} & 0 \\ 0 & 3 \end{bmatrix} \begin{bmatrix} 0 \\ 1 \end{bmatrix} = \begin{bmatrix} 0 \\ 3 \end{bmatrix} = 3 \begin{bmatrix} 0 \\ 1 \end{bmatrix}.$$

In other words, the vectors $\begin{bmatrix} 1 \\ 0 \end{bmatrix}$ and $\begin{bmatrix} 0 \\ 1 \end{bmatrix}$ are eigenvectors for the matrix associated with f_3 with eigenvalues $1/2$ and 3 respectively.

Generalizing this to any dynamical system $(\mathbb{R}^2, f)$ where f is a linear map, we note that if the matrix A representing f has an eigenvector $\vec{v} = \begin{bmatrix} v_1 \\ v_2 \end{bmatrix}$ with eigenvalue λ then for any $t \in \mathbb{R}$ and $k \in \mathbb{N}$,

$$f^k(tv_1, tv_2) = (\lambda^k tv_1, \lambda^k tv_2)$$

(Exercise 3a). That is, the line through the origin containing (v_1, v_2) is invariant under f and points on the line have iterates that approach the origin if $|\lambda| < 1$ and move away from the origin if $|\lambda| > 1$. The situation is harder to understand if $|\lambda| = 1$ (Exercise 4).

If, as in the examples, the linear map f has distinct eigenvalues λ and μ with eigenvectors $\vec{v} = \begin{bmatrix} v_1 \\ v_2 \end{bmatrix}$ and $\vec{w} = \begin{bmatrix} w_1 \\ w_2 \end{bmatrix}$ respectively, then we can describe the behavior of f on any initial condition by using the fact that $\vec{v}$ and $\vec{w}$ must form a basis of $\mathbb{R}^2$ (Exercise 2c). This means that given any $\begin{bmatrix} x \\ y \end{bmatrix}$ we know there exists constants c_1, c_2 so that

$$\begin{bmatrix} x \\ y \end{bmatrix} = c_1 \begin{bmatrix} v_1 \\ v_2 \end{bmatrix} + c_2 \begin{bmatrix} w_1 \\ w_2 \end{bmatrix}.$$

Using the linearity of matrix multiplication and the fact that $\vec{v}$ and $\vec{w}$ are eigenvectors we have

$$A \begin{bmatrix} x \\ y \end{bmatrix} = c_1 A \begin{bmatrix} v_1 \\ v_2 \end{bmatrix} + c_2 A \begin{bmatrix} w_1 \\ w_2 \end{bmatrix} = c_1 \lambda \begin{bmatrix} v_1 \\ v_2 \end{bmatrix} + c_2 \mu \begin{bmatrix} w_1 \\ w_2 \end{bmatrix}.$$

So $f(x, y) = c_1 \lambda (v_1, v_2) + c_2 \mu (w_1, w_2)$ and an induction argument shows that for all $k \in \mathbb{N}$,

$$f^k(x, y) = c_1 \lambda^k (v_1, v_2) + c_2 \mu^k (u_1, u_2)$$

(Exercise 3b).

Let $\mathcal{L}_1$ denote the line through the origin parallel to the eigenvector $\vec{v}$ and $\mathcal{L}_2$ the line through the origin parallel to the eigenvector $\vec{w}$. $\mathcal{L}_1$ and $\mathcal{L}_2$ are invariant under the action of A regardless of the values of λ and μ. However, the behavior of iterates of initial conditions that lie on them depend on λ and μ. In particular, if $|\lambda|, |\mu| < 1$, then clearly $\lim_{k \to \infty} f^k(x, y) = (0, 0)$ for all (x, y) on $\mathcal{L}_1$ and $\mathcal{L}_2$ and, by our argument above, for all $(x, y) \in \mathbb{R}^2$. Similarly, if $|\lambda|, |\mu| > 1$ then the iterates $f^k(x, y)$ grow without bound for all $(x, y) \in \mathbb{R}^2$. When $0 \leq |\lambda| < 1 < |\mu|$, the iterates of initial conditions on $\mathcal{L}_1$ converge to the origin, whereas the iterates of initial conditions on $\mathcal{L}_2$ move away from the origin. Thus the iterates of any $(x, y) \in \mathbb{R}^2$ not on $\mathcal{L}_1$ grow without bound. In this case the fixed point $(0, 0)$ is called a **saddle point**. These observations are summarized in the next theorem.

Theorem 9.4. *Let λ and μ be two distinct real-valued eigenvalues associated to the linear map f.*

1. *If $0 \leq |\lambda|, |\mu| < 1$, then $(0, 0)$ is an attracting fixed point.*
2. *If $|\lambda|, |\mu| > 1$, then $(0, 0)$ is a repelling fixed point.*
3. *If $0 \leq |\lambda| < 1 < |\mu|$, then $(0, 0)$ is a saddle point.*

The theorem does not apply if either $|\lambda| = 1$ or $|\mu| = 1$; in these cases many behaviors are possible (Exercise 4).

To illustrate these ideas, let us consider the map $g(x, y) = (y/4, -5x + 3y)$. You will show in Exercise 1c that the vectors $\begin{bmatrix} 1 \\ 2 \end{bmatrix}$ and $\begin{bmatrix} 1 \\ 10 \end{bmatrix}$ are eigenvectors for g with associated eigenvalues $1/2$ and $5/2$ respectively. From Theorem 9.4, we know that the fixed point $(0, 0)$ will be a saddle point. There will be two invariant lines, namely

$$\mathcal{L}_1 = \{(t, 2t) \mid t \in \mathbb{R}\} \text{ and } \mathcal{L}_2 = \{(t, 10t) \mid t \in \mathbb{R}\},$$

and the iterates of initial conditions on $\mathcal{L}_1$ will approach the fixed point $(0, 0)$ while the iterates of initial conditions on $\mathcal{L}_2$ will grow without bound. An initial condition (x_0, y_0) not on either of the invariant lines can be written as $c_1(1, 2)+c_2(1, 10)$ for some $c_1, c_2 \in \mathbb{R}$, and the iterates of (x_0, y_0) can be written as

$$g^k(x_0, y_0) = g^k(c_1(1, 2) + c_2(1, 10)) = c_1 g^k(1, 2) + c_2 g^k(1, 10).$$

Clearly the $c_1 g^k(1, 2)$ term of the iterates will approach zero while the $c_2 g^k(1, 10)$ term will grow without bound, so $g^k(x_0, y_0)$ will also grow without bound with iterates approaching the line $\mathcal{L}_2$. This is illustrated in Figure 9.2.

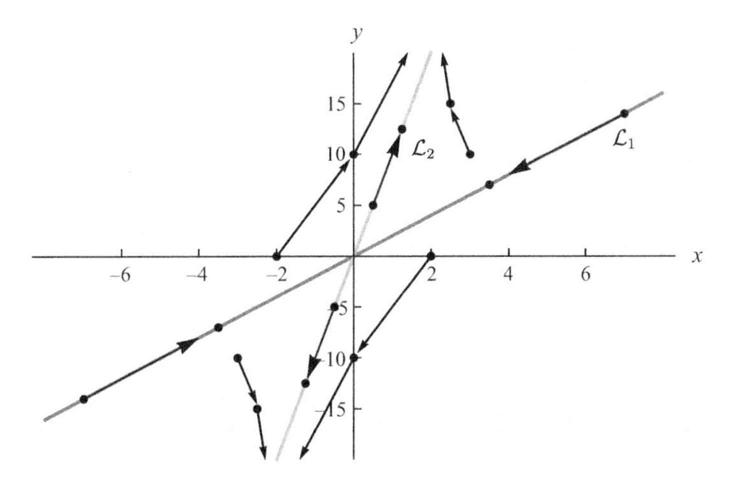

Figure 9.2. Invariant lines for g and the iterates of several initial conditions.

Of course, not all maps from $\mathbb{R}^2$ to $\mathbb{R}^2$ are linear. For instance, the map

$$h(x, y) = \left(x^2 + \frac{1}{2}x, y^3 + \frac{1}{3}y\right)$$

is not linear, but it still has $(0, 0)$ as a fixed point. In Figure 9.3, we iterate a point that is close to the origin to better understand the nature of this fixed point.

Figure 9.3 looks similar to Figure 9.1, which illustrates the iterates of the map f_1. This is not surprising: for (x, y) near $(0, 0)$, the x^2 and y^3 terms in h will be negligible and the action of h will be mainly determined by the terms $x/2$ and $y/3$.

We further support our observation that h behaves like f_1 near $(0, 0)$ by calculating the derivative at $(0, 0)$ of the two component functions of h. This will tell us the rate at which the function h is changing near $(0, 0)$ in each of the component directions. The derivative of $x^2 + x/2$ with respect to x at $(0, 0)$ is $1/2$ and thus the first coordinate of h changes in

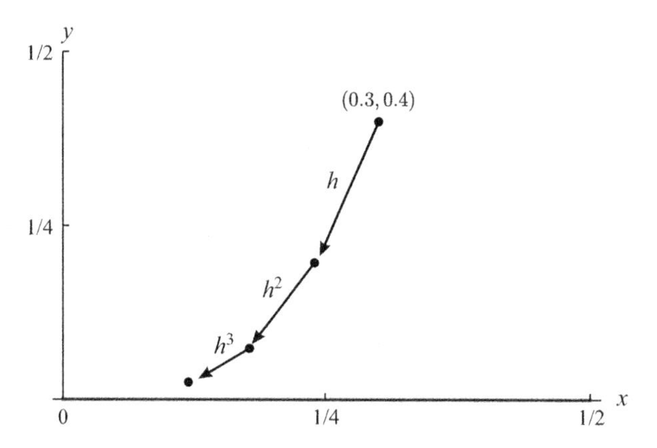

Figure 9.3. Iterates of the point $(0.3, 0.4)$ under the map h.

the x-direction in the same way as the first coordinate of f_1. Similarly, since the derivative of $y^3 + y/3$ with respect to y at $(0, 0)$ is $1/3$, the second coordinate of h changes in the y-direction in the same way as the second coordinate of f_1.

It turns out that any map $h : \mathbb{R}^2 \to \mathbb{R}^2$ behaves like a linear map near its fixed points. We can use the derivatives of the component functions to find the approximating linear map as follows:

Definition 9.5. *Let $h : \mathbb{R}^2 \to \mathbb{R}^2$ be given by $h(x, y) = (u(x, y), v(x, y))$, and let (p, q) be a fixed point for h. Then the* **Jacobian matrix** *of h at the fixed point (p, q) is the matrix of partial derivatives*

$$\begin{bmatrix} \frac{\partial u}{\partial x}(p,q) & \frac{\partial u}{\partial y}(p,q) \\ \frac{\partial v}{\partial x}(p,q) & \frac{\partial v}{\partial y}(p,q) \end{bmatrix}.$$

Let f be the linear map defined by the Jacobean matrix of h at the fixed point (p, q). It has a fixed point at $(0, 0)$ and its behavior near the origin closely approximates the behavior of h near its fixed point (p, q). Thus the nature of a fixed point (p, q) for an arbitrary map $h : \mathbb{R}^2 \to \mathbb{R}^2$ is the same as the nature of the fixed point $(0, 0)$ of this associated linear map f. Using linear maps in this way to understand the local behavior of nonlinear maps is one of the powerful applications of the ideas in this module.

Exercises

1. For the following linear maps, find and classify the fixed points and describe any invariant lines.

   (a) The linear map f defined in the exploration.

   (b) $h(x, y) = (x/12 - 5y/12, -5x/12 + y/12)$

   (c) $g(x, y) = (y/4, -5x + 3y)$

   (d) $k(x, y) = (-2x - y, 2x - 5y)$

2. Let $f : \mathbb{R}^2 \to \mathbb{R}^2$ be represented by the matrix $\begin{bmatrix} a & b \\ c & d \end{bmatrix}$.

(a) Show that λ is an eigenvalue of f exactly when $(a - \lambda)(d - \lambda) - bc = 0$.

(b) Show that if $\begin{bmatrix} x \\ y \end{bmatrix}$ is an eigenvector for the eigenvalue λ, then for any nonzero $t \in \mathbb{R}$, $\begin{bmatrix} tx \\ ty \end{bmatrix}$ is also an eigenvector for the eigenvalue λ.

(c) Show that the eigenvectors associated with two distinct, nonzero eigenvalues λ and μ cannot be scalar multiples of each other, and thus will give a basis for $\mathbb{R}^2$.

3. Let $f : \mathbb{R}^2 \to \mathbb{R}^2$ be represented by the matrix $\begin{bmatrix} a & b \\ c & d \end{bmatrix}$.

   Assume that $\begin{bmatrix} v_1 \\ v_2 \end{bmatrix}$ and $\begin{bmatrix} u_1 \\ u_2 \end{bmatrix}$ are eigenvectors associated with distinct eigenvalues λ and μ respectively.

   (a) Use induction to show that for any $t \in \mathbb{R}$ and $k \in \mathbb{N}$,
   $$f^k(tv_1, tv_2) = (\lambda^k tv_1, \lambda^k tv_2).$$

   (b) Let $(x, y) = c_1(v_1, v_2) + c_2(u_1, u_2)$. Use induction to show that
   $$f^k(x, y) = c_1\lambda^k(v_1, v_2) + c_2\mu^k(u_1, u_2).$$

4. Let $f : \mathbb{R}^2 \to \mathbb{R}^2$ be represented by the matrix $\begin{bmatrix} a & b \\ c & d \end{bmatrix}$.

   (a) Prove that f has a nonzero fixed point if and only if f has an eigenvalue $\lambda = 1$.

   (b) Find conditions on the constants a, b, c, and d for which f will have an eigenvalue $\lambda = 1$.

   (c) Choose an example of a linear map having an eigenvalue $\lambda = 1$ and determine the nature of its fixed points.

5. In this problem, we will consider a linear map $f(x, y) = (y, -x)$ which does not have real-valued eigenvalues.

   (a) Find the complex-valued eigenvalues for f.

   (b) Explain why f has no invariant lines through the origin.

   (c) Show that the distance between $f(x, y)$ and the origin is the same as the distance between (x, y) and the origin.

   (d) Show that any circle of fixed radius centered at the origin is an invariant set for f.

   (e) What can you say about the nature of the fixed point $(0, 0)$ for this example?

6. (a) Find and classify the fixed points of $h_1(x, y) = \left(y + x/2,\ y^2 + x\right)$.

   (b) Find and classify the fixed points of $h_2(x, y) = (-xy - 2x,\ 2xy - 3y)$.

   (c) For each of the functions in (a) and (b), iterate a variety of initial values near the fixed points to confirm their nature.

Project

Consider the family of maps $T_{ab}(x, y) = (a - by - x^2, 2y - b)$. By studying the fixed points and the nature of these fixed points, describe any bifurcations that occur as the parameters a and b vary. Find a pictorial way to demonstrate your results and discuss ways to further refine your picture.

Module 10

Iterated Function Systems

Exploration

Consider the triangle in the plane with vertices $(0, 0)$, $(1, 0)$, and $(0, 1)$. Choose an initial condition x_0 in the triangle and roll a die (or use a random number generator). If the die lands on 1 or 2, move half of the distance between x_0 and $(0, 0)$ towards $(0, 0)$; if the die lands on 3 or 4, move half of the distance between x_0 and $(1, 0)$ towards $(1, 0)$; if the die lands on 5 or 6, move half of the distance between x_0 and $(0, 1)$ towards $(0, 1)$.

What can you say about the long term behavior of the iterates for different values of x_0? How do the iterates move about the triangle? Is it possible to make predictions about their future locations?

Exposition

There are some physical systems whose evolution over time cannot be described by a single, deterministic rule. Instead, the rule governing the behavior of a system at a particular time might be random. For example, the growth rate of the bacteria we encountered in Module 1 might depend on the temperature of the room in which the petri dish is kept, and the temperature might depend on an erratic heating system which randomly works better on some days than others. In this module, we will study a class of systems that have an element of randomness. These systems are referred to as iterated function systems.

Definition 10.1. *Let $\{p_1, p_2, \ldots, p_k\}$ be points in $\mathbb{R}^n$ and let $0 < \alpha < 1$. For any $x \in \mathbb{R}^n$ and for each $1 \leq i \leq k$, let*

$$F_i(x) = \alpha(x - p_i) + p_i = \alpha x + (1 - \alpha)p_i.$$

We call the collection $\mathcal{F} = \{F_1, F_2, \ldots, F_k\}$ an **iterated function system** *or* **IFS.**

The two-dimensional IFS that was the subject of the exploration is given by

$$\{p_1, p_2, p_3\} = \{(0,0), (1,0), (0,1)\} \subseteq \mathbb{R}^2 \quad \text{and} \quad \alpha = \frac{1}{2}$$

with

$$F_i(x) = \frac{1}{2}(x - p_i) + p_i = \frac{1}{2}x + \frac{1}{2}p_i$$

for each $x \in \mathbb{R}^2$ and $1 \leq i \leq 3$.

In this module, we will study iterated function systems by considering two examples. We begin with a one-dimensional example given by

$$\{p_1, p_2\} = \{0, 1\} \subseteq \mathbb{R} \quad \text{and} \quad \alpha = \frac{1}{3}.$$

Then

$$F_i(x) = \frac{1}{3}(x - p_i) + p_i = \begin{cases} \frac{1}{3}x & \text{if } i = 1 \\ \frac{1}{3}x + \frac{2}{3} & \text{if } i = 2. \end{cases}$$

Like all iterated function systems, this IFS differs from the dynamical systems that we have studied up to this point in that the image of an initial condition under the IFS is not determined. Given an initial condition x_0, its first iterate under the IFS could be $F_1(x_0) = x_0/3$ or $F_2(x_0) = x_0/3 + 2/3$. Then, depending on the sequence of F_1 and F_2 used, there are four equally likely possibilities for its second iterate. The possibilities are:

$$\begin{aligned} F_1(F_1(x_0)) &= 1/3\,(x_0/3) & \qquad F_1(F_2(x_0)) &= 1/3\,(x_0/3 + 2/3) \\ F_2(F_1(x_0)) &= 1/3\,(x_0/3) + 2/3 & \qquad F_2(F_2(x_0)) &= 1/3\,(x_0/3 + 2/3) + 2/3. \end{aligned}$$

Continuing in this way, there are eight possibilities for the third iterate of x_0, and so on. Because of this, we need to rethink our notion of an orbit. Each infinite sequence of 1s and 2s determines a different set of iterates, each of which is a possible orbit of x_0. In general, we have the definition:

Definition 10.2. *Let $\mathcal{F} = \{F_1, F_2, \ldots, F_k\}$ be an IFS and let x_0 be an initial condition. Given a sequence $i_1, i_2, i_3, \ldots$ where $i_j \in \{1, 2, 3, \ldots, k\}$ for all $j \in \mathbb{N}$, the set of iterates*

$$x_0,\ F_{i_1}(x_0),\ F_{i_2}\left(F_{i_1}(x_0)\right),\ F_{i_3}\left(F_{i_2}\left(F_{i_1}(x_0)\right)\right), \ldots$$

is an **IFS orbit** *of x_0.*

In our study of dynamical systems up to this point, we have been interested in the long term behavior of orbits of initial conditions. Because an initial condition has multiple IFS orbits, there can be more than one possibility for the long term behavior. In our example, for any initial condition x_0, the sequence $1, 1, 1, \ldots$ yields the IFS orbit x_0, $(1/3)x_0$, $(1/3)^2 x_0$, $(1/3)^3 x_0$, $\ldots$, and we can see that the iterates tend to 0. On the other hand, the IFS orbit determined by the sequence $2, 2, 2, \ldots$ of the same initial condition is

$$x_0,\ \left(\frac{1}{3}\right)x_0+\frac{2}{3},\ \left(\frac{1}{3}\right)^2 x_0+\left(\frac{1}{3}\right)\frac{2}{3}+\frac{2}{3},\ \left(\frac{1}{3}\right)^3 x_0+\left(\frac{1}{3}\right)^2\frac{2}{3}+\left(\frac{1}{3}\right)\frac{2}{3}+\frac{2}{3},\ \ldots .$$

In this case, the iterates tend to

$$\lim_{m\to\infty}\left(\frac{1}{3}\right)^m x_0 + \frac{2}{3}\sum_{i=0}^{m-1}\left(\frac{1}{3}\right)^i = 1.$$

Thus, one IFS orbit of x_0 approaches 0 while another approaches 1. By choosing different sequences of 1s and 2s, we might expect to get many other possible long term behaviors of the IFS orbits of x_0.

In order to understand all the possible long term behaviors of the IFS orbits of x_0, we will need a good understanding of a general IFS orbit. Choose an arbitrary sequence $i_1, i_2, i_3, \ldots$ with $i_j \in \{1, 2\}$ for each $j \in \mathbb{N}$. We can use induction (Exercise 2) to show that the mth iterate in this IFS orbit is of the form

$$\frac{x_0}{3^m} + \sum_{k=1}^{m}\frac{2c_{m-k+1}}{3^k} \tag{10.1}$$

where

$$c_j = \begin{cases} 0 & \text{if } i_j = 1 \text{ and thus } F_1 \text{ is used for the } j^{th} \text{ iteration} \\ 1 & \text{if } i_j = 2 \text{ and thus } F_2 \text{ is used for the } j^{th} \text{ iteration.} \end{cases}$$

If we take the limit as m increases to infinity we get

$$\sum_{k=1}^{\infty}\frac{2c_{m-k+1}}{3^k}.$$

The first thing we notice about this limit is that it does not depend on x_0. This leads to the rather surprising observation that the long term behavior of an IFS orbit does not depend on the initial condition but rather it is completely determined by the sequence $i_1, i_2, \ldots$. The second observation is that the limit is a point whose ternary expansion consists of only 0s and 2s, and therefore (Exercise 5, Module 4) it is a point that lies in the Cantor set Γ. Furthermore, by varying $i_1, i_2, \ldots$ we get every point in Γ as the limit of some IFS orbit of x_0.

In summary, for this example we have found that the set Γ is exactly the set of limit points of all IFS orbits for any initial condition. The set Γ is called the **attracting set**, or **attractor**, for the IFS. The following theorem tells us that every IFS has an attractor.

Theorem 10.3. *Let $\mathcal{F} = \{F_1, F_2, \ldots, F_k\}$ be an IFS. There is a unique closed and bounded set $A \subseteq \mathbb{R}^n$ with the properties*

1. *The limit of any IFS orbit of any initial condition will be in the set A; and*
2. *For any point $a \in A$ and any initial condition $x_0 \in \mathbb{R}^n$, there is some sequence $i_1, i_2, i_3, \ldots$ with $i_j \in \{1, 2, 3, \ldots, k\}$, $j \in \mathbb{N}$, for which the limit of the IFS orbit of x_0 determined by the sequence is the point a.*

The proof of this and other theorems in this module rely on the contraction mapping theorem and ideas in topology that take us too far afield, and so we omit them. The interested reader is directed to the works of, for example, Barnsley [Ba], Hutchinson [H], and Falconer [F].

Returning to our one-dimensional example, we now consider an initial condition $x_0 \in \Gamma$. Then $x_0 = \sum_{j=1}^{\infty} \frac{a_j}{3^j}$ where each a_j is either 0 or 2. We notice that

$$F_1(x_0) = \frac{1}{3}\left(\sum_{j=1}^{\infty} \frac{a_j}{3^j}\right) = \sum_{j=1}^{\infty} \frac{a_j}{3^{j+1}},$$

which can be rewritten $\sum_{j=1}^{\infty} \frac{b_j}{3^j}$ where $b_1 = 0$ and $b_j = a_{j-1}$ for $j \geq 2$. So, $F_1(x_0) \in \Gamma$. Similarly

$$F_2(x_0) = \frac{1}{3}\left(\sum_{j=1}^{\infty} \frac{a_j}{3^j}\right) + \frac{2}{3} = \sum_{j=1}^{\infty} \frac{a_j}{3^{j+1}} + \frac{2}{3},$$

which can be rewritten $\sum_{j=1}^{\infty} \frac{b_j}{3^j}$ where $b_1 = 2$ and $b_j = a_{j-1}$ for $j \geq 2$. So, $F_2(x_0) \in \Gamma$ as well. Since x_0 was an arbitrary point in Γ, this shows us that for any $x_0 \in \Gamma$, every IFS orbit of x_0 stays in Γ.

More generally, let $\mathcal{F} = \{F_1, F_2, \ldots, F_k\}$ be an arbitrary IFS and let S be a subset of $\mathbb{R}^n$. Let $\mathcal{F}(S)$ denote all possible images of all possible initial conditions in S. That is,

$$\mathcal{F}(S) = \bigcup_{x_0 \in S} \{\, F_1(x_0), F_2(x_0), \ldots, F_k(x_0) \,\}.$$

The preceding paragraph tells us that for our example, $\mathcal{F}(\Gamma) \subseteq \Gamma$. It is not difficult to show that $\Gamma \subseteq \mathcal{F}(\Gamma)$ as well (Exercise 4), and thus $\mathcal{F}(\Gamma) = \Gamma$. We say that Γ is an **invariant set** for the IFS. It is not hard to see that it is the only nontrivial invariant set for the IFS. The following theorem generalizes these observations to any IFS.

Theorem 10.4. *Let $\mathcal{F} = \{F_1, F_2, \ldots, F_k\}$ be an IFS and A be a closed and bounded subset of $\mathbb{R}^n$. Then A is an attracting set for $\mathcal{F}$ if and only if $\mathcal{F}(A) = A$.*

In the next example we will see how to put Theorem 10.4 to use in determining the attracting set of an IFS. Consider the two-dimensional IFS given by

$$\{p_1, p_2, p_3, p_4\} = \{(0,0), (1,0), (0,1), (1,1)\} \subseteq \mathbb{R}^2 \text{ and } \alpha = \frac{1}{3}.$$

Then

$$F_i(x) = \frac{1}{3}(x - p_i) + p_i = \frac{1}{3}x + \frac{2}{3}p_i.$$

Denote the attracting set of this IFS by Λ. In order to find Λ using Theorem 10.4, we will look for a set whose image under F_1, F_2, F_3, and F_4 is invariant. Since $F_i(p_i) = p_i$, we know that the p_is will be in Λ, and thus we take as our first approximation to Λ the set S where S is the unit square with lower left corner at the origin. Each F_i maps S onto a square of side length $\frac{1}{3}$ with one of its four corners at p_i, so $\mathcal{F}(S)$ is the union of the four squares shaded in Figure 10.1. This tells us that the whole of S is not the invariant set for the IFS since $\mathcal{F}(S)$ is a proper subset of S. However, the initial conditions in the four squares of $\mathcal{F}(S)$ have the property that they remain in $\mathcal{F}(S)$ after one iteration of any of the F_is in the IFS, so these initial conditions are possible elements of Λ.

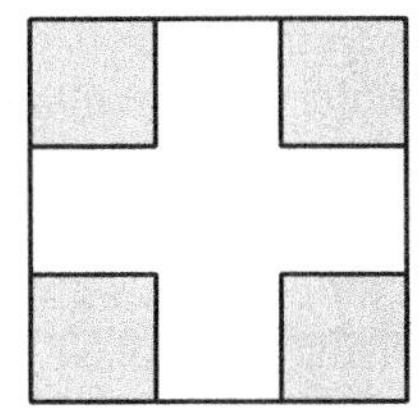

Figure 10.1. The image $\mathcal{F}(S)$ of S under one iteration of functions in $\mathcal{F}$.

The next iteration maps each of the four shaded squares in Figure 10.1 to four new squares, each a third of the previous size, as illustrated in Figure 10.2. The initial conditions in the sixteen smaller squares making up $\mathcal{F}(\mathcal{F}(S))$ have the property that they remain in $\mathcal{F}(\mathcal{F}(S))$ after two iterations of any of the F_is in the IFS. Thus the initial conditions in $\mathcal{F}(\mathcal{F}(S))$ are possible elements of Λ.

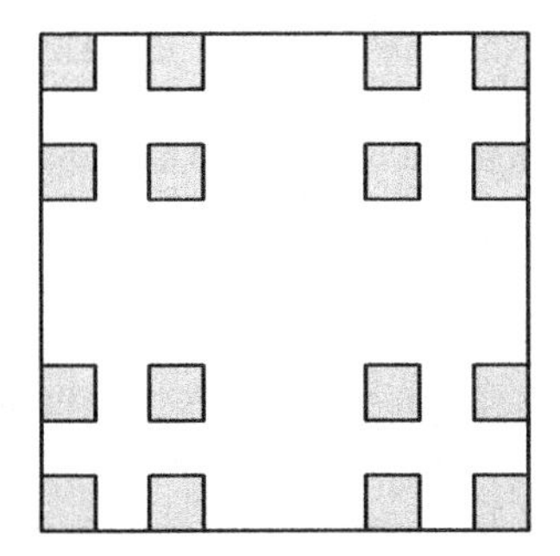

Figure 10.2. The image $\mathcal{F}(\mathcal{F}(S))$ of S after two iterations of functions in $\mathcal{F}$.

We can continue in this way and at the mth stage we will have a set of initial conditions that remain in the set after all possible combinations of m functions from the IFS. You will finish the construction of the set Λ in Exercise 5. An illustration of the set Λ is shown in Figure 10.3.

The set Λ shares an important characteristic with the Cantor set Γ: both sets are self-similar. There are various definitions of self-similar, but one often used is:

Definition 10.5. *A set S is* **self-similar** *if it can be subdivided into a finite number of subsets, each of which can be magnified to give S.*

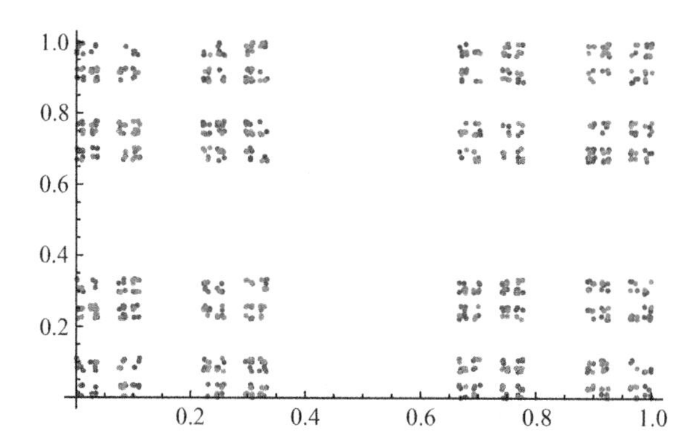

Figure 10.3. The attractor Λ for $\mathcal{F}$.

Informally, a portion of a self-similar set resembles the whole set after magnification. For example, if we magnify the portion of Γ contained in $[0, 1/3]$ or $[2/3, 1]$ by a factor of three, we will obtain all of Γ. Similarly, if we magnify the intersection of Λ with any of the four boxes in Figure 10.1 by a factor of three, we will obtain the entire set Λ. There are usually many choices for the subdivision mentioned in Definition 10.5. For instance, the intersection of Γ with $[0, 1/9]$, $[2/9, 1/3]$, $[2/3, 7/9]$, or $[8/9, 1]$ can all be magnified by a factor of nine to obtain Γ. The intersection of Λ with the sixteen boxes shown in Figure 10.2 yields sixteen subsets, each of which can be magnified by a factor of nine to give back the entire attractor.

The following theorem tells us that it is no coincidence that the attracting sets for our two examples of iterated function systems are both self-similar.

Theorem 10.6. *Let $\mathcal{F} = \{F_1, F_2, \ldots, F_k\}$ be an IFS. The attracting set for $\mathcal{F}$ is self-similar.*

The sets Λ and Γ are both examples of fractal sets. Definitions of fractal sets vary, but one of the defining characteristics of a fractal is self-similarity. Another defining characteristic of a fractal involves the dimension of the set. The box dimension was introduced in the project of Module 4. We will use a different definition of dimension here, and in Exercise 9 you will show that the two definitions are equivalent.

Definition 10.7. *Let S be a self-similar set that can be divided into K subsets, each of which can be magnified by a factor m to give S. The* **fractal dimension** *of S is the solution $d \in \mathbb{R}$ of the equation $m^d = K$.*

This definition of fractal dimension is consistent with our intuition for non-fractal sets. For example, a line segment can be divided in half to give two similar subsets, each of which gives the whole line segment when magnified by a factor of two. So in this case, $K = 2$ and $m = 2$ and the solution to $2^d = 2$ is one.

Now let us use Definition 10.7 to find the fractal dimension of Γ. We note that $\Gamma \cap [0, 1/3]$ and $\Gamma \cap [2/3, 1]$ give two subsets of Γ. Each can be magnified by a factor of three to give Γ. Thus $K = 2$ and $m = 3$ and solving the equation $3^d = 2$ gives

$$d = \frac{\ln(2)}{\ln(3)} \approx 0.63.$$

We noted earlier that the way of dividing a self-similar set into subsets need not be unique; however, any choice for the subsets will yield the same fractal dimension. For example,

$$\Gamma \cap \left[0, \frac{1}{9}\right], \quad \Gamma \cap \left[\frac{2}{9}, \frac{1}{3}\right], \quad \Gamma \cap \left[\frac{2}{3}, \frac{7}{9}\right], \quad \Gamma \cap \left[\frac{8}{9}, 1\right]$$

are four subsets of Γ. Each can be magnified by a factor of nine to give Γ. Thus $K = 4$ and $m = 9$ and solving the equation $9^d = 4$ gives

$$d = \frac{\ln(4)}{\ln(9)} = \frac{\ln(2^2)}{\ln(3^2)} = \frac{2\ln(2)}{2\ln(3)} \approx 0.63.$$

We are now ready to state one common definition of a fractal set.

Definition 10.8. *A set S is a* **fractal** *if it is self-similar and has non-integer fractal dimension.*

Fractals are interesting sets, and they have been studied in a variety of contexts. Graphical illustrations of fractal sets can be stunningly complex and beautiful. After an introduction to complex dynamical systems in Module 11, we will consider in Module 12 two well-known fractals that arise in the study of complex dynamical systems, namely the Julia and Mandelbrot sets.

Exercises

1. Let $F_2(x) = (1/3)x + 2/3$.

   (a) Find the first four iterates of $x_0 = 1/2$ by F_2.

   (b) Find a formula for $F_2^m(1/2)$.

   (c) Find $\lim_{m\to\infty} F_2^m(1/2)$.

2. Let $\mathcal{F}$ be the one-dimensional IFS defined in the exposition by

$$\{p_1, p_2\} = \{0, 1\} \subset \mathbb{R} \text{ and } F_i(x) = \frac{1}{3}(x - p_i) + p_i$$

   for $x \in \mathbb{R}$ and $1 \le i \le 2$. Use induction to prove that the mth iterate of an IFS orbit of x_0 is

$$\frac{x_0}{3^m} + \sum_{i=1}^{m} \frac{2c_{m-i+1}}{3^i} = \frac{x_0}{3^m} + \frac{2c_m}{3} + \frac{2c_{m-1}}{3^2} + \cdots + \frac{2c_1}{3^m},$$

   where $c_j = 0$ when $i_j = 1$ and $c_j = 1$ when $i_j = 2$.

3. Let $\mathcal{F}$ be the one-dimensional IFS defined in the exposition and in Exercise 2, and let

$$\Gamma = \bigcap_{n\in\mathbb{N}} \Gamma_n$$

   be the Cantor set as defined in Module 4. Show that $\mathcal{F}(\Gamma_1) = \Gamma_2$. Explain why in general $\mathcal{F}(\Gamma_n) = \Gamma_{n+1}$.

4. Let $\mathcal{F}$ be the one-dimensional IFS defined in the exposition and in Exercise 2. Show that $\Gamma \subseteq \mathcal{F}(\Gamma)$.

5. Let $\mathcal{F}$ now be the two-dimensional IFS defined in the exposition by

$$\{p_1, p_2, p_3, p_4\} = \{(0,0), (1,0), (0,1), (1,1)\} \subset \mathbb{R}^2 \text{ and } F_i(x) = \frac{1}{3}(x - p_i) + p_i$$

for $x \in \mathbb{R}^2$ and $1 \leq i \leq 4$. Let Λ be its attracting set.

   (a) Find sets S_n such that $\Lambda = \bigcap_{n \in \mathbb{N}} S_n$.
   (Hint: This is similar to what we did in Module 4 for the Cantor set.)

   (b) A set $S \subset \mathbb{R}^2$ is **totally disconnected** if it contains no open balls. In other words, for every $(p, q) \in \mathbb{R}^2$ and every $\epsilon > 0$, the set

$$\{(x, y) : d((x, y), (p, q)) < \epsilon\}$$

   is not contained in S. Show that Λ is totally disconnected.

   (c) Show that Λ is uncountable.

6. Find the attracting set for the IFS given in the exploration.

7. Create a two-dimensional IFS and find its attractor.

8. Motivated by the description in the exploration, Definition 10.1 requires α to be in the interval $(0, 1)$. Create examples of iterated function systems with α outside this interval and discuss the possible long term behavior of initial conditions.

9. Let $S \subset \mathbb{R}$ be a self-similar set that can be divided into K subsets, each of which can be magnified by a factor m to give S. Show that the fractal dimension of S is the same as its box counting dimension as defined in the Module 4 project.

10. Find the fractal dimension of each of the following sets. Determine if each set is (i) self similar, and (ii) a fractal.

   (a) a square

   (b) the attracting set for the IFS given in the exploration

   (c) the set Λ

   (d) the attracting set of the IFS defined in Exercise 7

11. Consider the following construction: begin with a line segment of length one. Construct an equilateral triangle on the middle third of the line and erase the base of the triangle. Repeat the process on the resulting four line segments (each of which has length one third.) Continue this process on each smaller line segment, ad infinitum, as shown in Figure 10.4. The limiting object is known as the Koch curve.

   Find its fractal dimension.

Figure 10.4. The first steps in creating the Koch curve.

12. Another interesting fractal is the Koch snowflake. To create it, begin with an equilateral triangle where each side has length one. Create a Koch curve on each of its sides; Figure 10.5 shows the first three steps. The resulting object is the Koch snowflake. Find the area enclosed by the Koch snowflake. What is the length of its boundary?

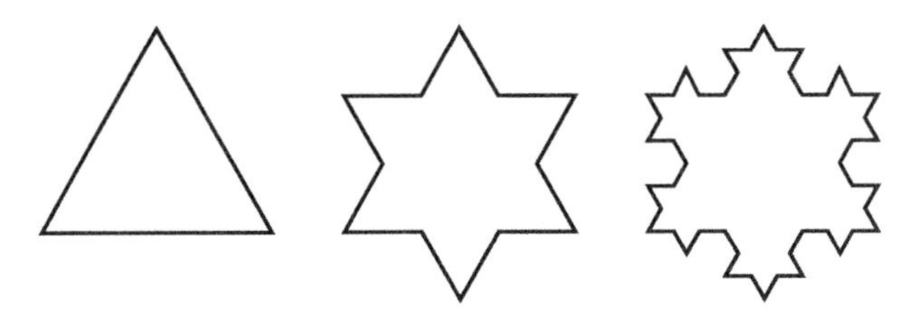

Figure 10.5. The first steps in creating the Koch snowflake.

Project

In this section we have defined iterated function systems via maps F_i of the form $F_i(x) = \alpha x + (1-\alpha)p_i$. By viewing points in $\mathbb{R}^n$ as vectors, we can rewrite the maps F_i using matrices:

$$F_i(\vec{x}) = A_i\,\vec{x} + \vec{b}_i$$

where $\vec{b}_i$ and $\vec{x}$ are in $\mathbb{R}^n$ and A_i is an $n \times n$ matrix. For example, the two-dimensional system described in the exploration can be expressed as

$$F_1(\vec{x}) = \begin{bmatrix} 1/2 & 0 \\ 0 & 1/2 \end{bmatrix}\vec{x} + \begin{bmatrix} 0 \\ 0 \end{bmatrix}, \quad F_2(\vec{x}) = \begin{bmatrix} 1/2 & 0 \\ 0 & 1/2 \end{bmatrix}\vec{x} + \begin{bmatrix} 1/2 \\ 0 \end{bmatrix}, \text{ and}$$

$$F_3(\vec{x}) = \begin{bmatrix} 1/2 & 0 \\ 0 & 1/2 \end{bmatrix}\vec{x} + \begin{bmatrix} 0 \\ 1/2 \end{bmatrix}.$$

1. Find matrices A_i and vectors $\vec{b}_i$ that express the two-dimensional system described in the exposition.

2. In the examples, all the A_i were identical diagonal matrices and all the $\vec{b}_i$ were in the unit square. In general, the A_i need not be identical or diagonal, and there can be any number of them. The $\vec{b}_i$ can be any point in $\mathbb{R}^n$. Use technology to generate a variety of two-dimensional examples and make conjectures about the effect of varying the A_i and the $\vec{b}_i$. In particular, conjecture under what conditions the IFS will have an attracting set.

Module 11

Dynamics in the Complex Plane

Exploration

Consider the function $f_1(z) = \dfrac{z+1}{3-z}$. Iterate f_1 for a variety of initial conditions in the complex plane $\mathbb{C}$ and try to find all possible long term behaviors.

Do the same for $f_2(z) = \dfrac{2z+4}{z-1}$.

Exposition

Complex numbers were alluded to as far back as the first century, but only became prominent in the sixteenth century when it was realized that they were necessary for finding roots of polynomials. Even though they were defined for purely theoretical reasons, they play an important role in many areas of mathematics, such as number theory, fluid mechanics, control theory, and signal analysis. In this module, we consider some of the dynamics that can occur when our phase space involves the complex plane. We focus this discussion on the family of rational functions called **fractional linear transformations** or **Möbius**[1] **transformations**. They are defined by

$$f(z) = \frac{az+b}{cz+d} \text{ where } ad - bc \neq 0.$$

Without the assumption that $ad - bc \neq 0$, the function f is a constant function (Exercise 1).

The function f is undefined at $z = -d/c$ and there are many initial conditions z_0 in the domain of f with the property that for some n, $f^n(z_0) = -d/c$. As you have seen in the exploration, this may make it difficult to identify the phase space of a dynamical system involving f. We will use a common technique from complex analysis to rectify this situation. We first recall that the distance between two complex numbers $z_1 = x + iy$ and $z_2 = a + ib$ is the same as the distance between them when they are considered as points (x, y) and (a, b) in the plane $\mathbb{R}^2$. With this notion of distance, the usual definitions of limits and continuity that we have been using for $\mathbb{R}$ and $\mathbb{R}^2$ extend readily to $\mathbb{C}$.

Note that

$$\lim_{z \to -\frac{d}{c}} |f(z)| = \infty.$$

Because of this, if ∞ were a complex number and we simply assigned $f(-d/c) = \infty$, then not only would f be defined on all of $\mathbb{C}$ but it would also be continuous at $-d/c$. However, assigning $f(-d/c) = \infty$ leads to a new problem when we want to continue iterating the initial condition $-d/c$: we need to assign a value to $f(\infty)$ in order for $f^2(-d/c)$ to be defined. As we did when defining $f(-d/c)$, we can check the behavior of f as $z \to \infty$ and we see that

$$\lim_{z \to \infty} f(z) = \frac{a}{c},$$

and thus for f to be continuous at ∞ we should define $f(\infty) = a/c$.

As every calculus student knows, infinity is an abstract concept and not an actual number, but the structure of the complex plane allows us to make sense of it as a value on what is called the extended complex plane:

Definition 11.1. *The* **extended complex plane**, *denoted by* $\mathbb{C}_\infty$, *is*

$$\mathbb{C}_\infty = \mathbb{C} \cup \{\infty\}.$$

One way of imagining this new space is to picture $\mathbb{C}$ as an infinitely large rectangle with a drawstring along its edges. When the drawstring is tightened, $\mathbb{C}$ is transformed from a flat plane into a sphere missing its north pole. The **point at infinity** is the north pole of

[1] August Möbius (1790–1868) was a German mathematician and astronomer who wrote many papers in geometry and in particular laid the groundwork for the development of projective geometry. However, he is perhaps best known for his work in topology and his discussion of the one-sided surface that is now called a Möbius band.

this sphere. The extended complex plane is also called the **Riemann[2] sphere** and there is a homeomorphism between $\mathbb{C}_\infty$ and the unit sphere centered at the origin (Exercise 10) that identifies the sphere (without the north pole) with $\mathbb{C}$.

Extending the domain of f to $\mathbb{C}_\infty$, we now give the formal definition of a Möbius transformation.

Definition 11.2. *A* **Möbius transformation** *is a function $f : \mathbb{C}_\infty \to \mathbb{C}_\infty$ of the form*

$$f(z) = \begin{cases} \frac{az+b}{cz+d} & \text{if } z \neq \infty, -\frac{d}{c} \\ \infty & \text{if } z = -\frac{d}{c} \\ \frac{a}{c} & \text{if } z = \infty \end{cases}$$

where $a, b, c, d \in \mathbb{C}$ with $ad - bc \neq 0$.

As discussed above, f is continuous at $z = -d/c$ and ∞. As a rational function, it is also continuous at all other points in its domain. Thus $f : \mathbb{C}_\infty \to \mathbb{C}_\infty$ is continuous. It is also one-to-one and onto and has an inverse that is itself a Möbius transformation (Exercise 2). We note that if $c = 0$, then d/c and a/c are undefined and Definition 11.2 reduces to

$$f(z) = \begin{cases} \frac{az+b}{d} & \text{if } z \neq \infty \\ \infty & \text{if } z = \infty. \end{cases}$$

In what follows, when we describe a Möbius transformation simply by $f(z) = \frac{az+b}{cz+d}$, the rest of the definition is implied.

Fixed points, periodic points, and cycles are defined for dynamical systems $(\mathbb{C}_\infty, f)$ as they were defined earlier for dynamical systems with real-valued phase spaces. If a fixed point p is in $\mathbb{C}$, then attracting and repelling can be readily adapted to this context using the notion of distance on $\mathbb{C}$. If $p = \infty$, we say that $z \in \mathbb{C}$ is close to p if its distance from the origin is large (see Module 1 Exercise 9), so we can also define these concepts for a fixed point at infinity.

With this in mind, we begin our study of systems $(\mathbb{C}_\infty, f)$ by asking the basic question about the number and nature of fixed points. As you will prove in Exercise 3, there are only two possibilities for the number of fixed points.

Theorem 11.3. *Let $f : \mathbb{C}_\infty \to \mathbb{C}_\infty$ be a Möbius transformation that is not the identity function. Then f has at least one and at most two fixed points.*

We first consider the case of a single fixed point. We begin with an example.

Example 11.4. *Let $g_1(z) = z + 1$.*

In this example, $a = b = 1$, $c = 0$, and $d = 1$. We see that ∞ is a fixed point of g_1. For any other $z \in \mathbb{C}_\infty$, the iterates of z are

$$z + 1, z + 2, z + 3, z + 4, \cdots,$$

[2] Georg Riemann (1826–1866) was a German mathematician who originally studied theology in order to become a pastor. He founded the field of Riemannian geometry, setting the stage for Einstein's general theory of relativity.

and $g_1^n(z) = z + n$ for all $n \in \mathbb{N}$. By using the closed form expression for $g_1^n(z)$, namely $g_1^n(z) = z + n$, we are able to quickly and completely characterize the dynamics of the function g_1. Clearly the iterates of all complex initial conditions tend towards $z = \infty$, and therefore $z = \infty$ is an attracting fixed point with basin of attraction $\mathbb{C}_\infty$.

Although this example is simple, it in fact captures the only type of dynamical behavior that a Möbius transformation with a single fixed point can exhibit. We will show this by proving that any Möbius transformation with a single fixed point is conjugate to $(\mathbb{C}_\infty, g_1)$. Consider a seemingly more complicated example with a single fixed point.

Example 11.5. *Let* $h_1(z) = \frac{2z-1}{z}$.

By solving $h_1(z) = z$, we find that the only fixed point is $z = 1$. Iterating a few different initial conditions yields some intuition that perhaps $z = 1$ is an attracting fixed point. However, unlike the previous example, there is no closed form expression for $h_1^n(z)$ that we can use to prove results about the nature of the fixed point.

Suppose we can show that there is a conjugacy $\psi : \mathbb{C}_\infty \to \mathbb{C}_\infty$ between $(\mathbb{C}_\infty, g_1)$ and $(\mathbb{C}_\infty, h_1)$. Since conjugacies preserve the number and nature of fixed points, this will show definitively that $z = 1$ is an attracting fixed point with basin of attraction $\mathbb{C}_\infty$. When showing two dynamical systems are conjugate, determining the conjugacy map can be difficult. To find such a map ψ in this case, we look for a function that maps the fixed point of h_1 to the fixed point of g_1. In other words we need $\psi(1) = \infty$, and thus candidates for ψ include rational functions with $z - 1$ in the denominator. We can try the simplest such example,

$$\psi(z) = \frac{1}{z-1},$$

and check to see if it is a conjugacy.

We first note that the map ψ is itself a Möbius transformation, and as such, it is one-to-one, onto, and continuous. Its inverse, $\psi^{-1}(z) = \frac{z+1}{z}$, is also a Möbius transformation and thus is continuous as well. So, ψ is a homeomorphism, and to show it is a conjugacy, we need only show that it commutes with h_1 and g_1.

First consider $\psi \circ h_1(z)$. We see that

$$\psi\left(h_1(z)\right) = \psi\left(\frac{2z-1}{z}\right) = \frac{1}{\frac{2z-1}{z} - 1} = \frac{z}{2z-1-z} = \frac{z}{z-1}.$$

Next consider $g_1 \circ \psi(z)$. We see that

$$g_1 \circ \psi(z) = g_1\left(\frac{1}{z-1}\right) = \frac{1}{z-1} + 1 = \frac{1+(z-1)}{z-1} = \frac{z}{z-1}.$$

So $\psi \circ h_1 = g_1 \circ \psi$ and thus $\psi : \mathbb{C}_\infty \to \mathbb{C}_\infty$ is a conjugacy as desired.

The argument used in Example 11.5 can be generalized to any Möbius transformation that has only one fixed point in order to prove the following theorem (Exercise 8).

Theorem 11.6. *Let* $f(z) = \frac{az+b}{cz+d}$ *be a Möbius transformation with exactly one fixed point* ζ. *Then the fixed point* ζ *is attracting with basin of attraction* $\mathbb{C}_\infty$.

Theorem 11.6 characterizes the dynamics of all Möbius transformations with a single fixed point. Next we turn our attention to the case of Möbius transformations with two fixed points. We begin with a simple example.

Example 11.7. *Let* $g_2(z) = 5z$.

This function is a Möbius transformation with $a = 5$, $b = c = 0$, and $d = 1$. It has two fixed points, 0 and ∞. As in Example 11.4, there is a closed form expression for the iterates of any $z \in \mathbb{C}_\infty$, namely $g_2^n(z) = 5^n z$. Thus, for $z \neq 0$, $g_2^n(z) \to \infty$, and we can completely characterize the dynamics of this example. It has one repelling fixed point, $z = 0$, and one attracting fixed point, $z = \infty$, with basin of attraction all nonzero points in $\mathbb{C}_\infty$.

Let us next consider a Möbius transformation that also has two fixed points, but is not as easily understood.

Example 11.8. *Let* $h_2(z) = \dfrac{3z-2}{-2z+3}$.

By setting $h_2(z) = z$, we can solve for the two fixed points of h_2, namely $z = 1$ and $z = -1$. As you might have guessed, $(\mathbb{C}_\infty, h_2)$ and $(\mathbb{C}_\infty, g_2)$ are conjugate. You will verify this fact in Exercise 4 and use it to show that $z = 1$ is a repelling fixed point for $(\mathbb{C}_\infty, h_2)$ and $z = -1$ is an attracting fixed point with basin of attraction all of $\mathbb{C}_\infty$ except $z = 1$.

The following theorem tells us how to construct a conjugacy between $(\mathbb{C}_\infty, h_2)$ and $(\mathbb{C}_\infty, g_2)$ and how to tell which fixed point of $(\mathbb{C}_\infty, h_2)$ will be attracting and which will be repelling.

Theorem 11.9. *Let* $f(z) = \frac{az+b}{cz+d}$ *be a Möbius transformation with two distinct fixed points,* ζ_1 *and* ζ_2. *Let*

$$\phi(z) = \frac{z-\zeta_1}{z-\zeta_2} \quad \textit{and} \quad \phi^{-1}(z) = \frac{-\zeta_2 z + \zeta_1}{-z+1}.$$

Then for some $k \in \mathbb{C}$ *and* $g(z) = kz$, $(\mathbb{C}_\infty, f)$ *and* $(\mathbb{C}_\infty, g)$ *are conjugate, and* ϕ *is the conjugacy between them. Furthermore,*

1. *if* $|k| > 1$, *then* ζ_1 *is repelling and* ζ_2 *is attracting with basin of attraction all points in* $\mathbb{C}_\infty$ *except* ζ_1

2. *if* $|k| < 1$, *then* ζ_2 *is repelling and* ζ_1 *is attracting with basin of attraction all points in* $\mathbb{C}_\infty$ *except* ζ_2.

If $(\mathbb{C}_\infty, f)$ is conjugate to $(\mathbb{C}_\infty, g)$ with $g(z) = kz$ and $|k| = 1$, the theorem does not apply and more investigation is needed to determine the nature of the fixed point. In Exercise 9 you will consider several examples of this type.

Möbius transformations have many other interesting properties. For example, they map lines and circles in the complex plane onto lines and circles, and they preserve the angle between curves. Interested readers can find proofs of these and other facts in any complex analysis book, including those by Beardon [Be] and Barnsley [Ba]. Of course, there are many functions defined on the complex plane that are not Möbius transformations. You will explore some in the project and the next module.

Exercises

1. Show that if $ad - bc = 0$ then $f(z) = \frac{az+b}{cz+d}$ is a constant function.

2. Show that a Möbius transformation f as given in Definition 11.2 is one-to-one and onto. Find the inverse f^{-1} of the function f.

3. Prove Theorem 11.3.

4. Let $(\mathbb{C}, g_2)$ and $(\mathbb{C}, h_2)$ be as defined in Examples 11.7 and 11.8, and define $\phi : \mathbb{C}_\infty \to \mathbb{C}_\infty$ as
$$\phi(z) = \frac{z-1}{z+1}.$$
   (a) Show that ϕ maps the fixed points of $(\mathbb{C}_\infty, h_2)$ to the fixed points of $(\mathbb{C}_\infty, g_2)$.

   (b) Show that ϕ is a conjugacy between $(\mathbb{C}_\infty, h_2)$ and $(\mathbb{C}_\infty, g_2)$.

5. Construct a conjugacy between $(\mathbb{C}_\infty, f_1)$ and $(\mathbb{C}_\infty, \tilde{g})$, where f_1 was defined in the exploration and $\tilde{g}$ is a Möbius transformation of the form $\tilde{g}(z) = z + \beta$. In other words, find $\beta \in \mathbb{C}$ such that $(\mathbb{C}_\infty, f_1)$ and $(\mathbb{C}_\infty, \tilde{g})$ are conjugate.

6. Construct a conjugacy between $(\mathbb{C}_\infty, f_2)$ and $(\mathbb{C}_\infty, g)$, where f_2 was defined in the exploration and g is a Möbius transformation of the form $g(z) = kz$. In other words, find $k \in \mathbb{C}$ such that $(\mathbb{C}_\infty, f_2)$ and $(\mathbb{C}_\infty, g)$ are conjugate.

7. For simplicity, the constants a, b, c, and d in the examples encountered thus far in the module have been real-valued, but in fact the constants can be complex-valued. Consider the Möbius transformation
$$f(z) = \frac{-iz}{(1-i)z + (-2+i)}$$
where $a = -i$, $b = 0$, $c = 1 - i$, and $d = -2 + i$. Verify that f has two fixed points, and then construct a conjugacy between $(\mathbb{C}_\infty, f)$ and $(\mathbb{C}_\infty, g)$ where $g(z) = kz$ for some $k \in \mathbb{C}$. Use this conjugacy to describe the long term behavior of initial conditions from $(\mathbb{C}_\infty, f)$.

8. In this problem we prove Theorem 11.6.

   (a) Let f be a Möbius transformation and assume ∞ is the only fixed point of f. Show that f must take the form $f(z) = z + \beta$ for some $\beta \in \mathbb{C}$.

   (b) Let $f(z) = \frac{az+b}{cz+d}$ be a Möbius transformation. Find conditions on a, b, c, and d such that f has exactly one fixed point, not equal to infinity, and give that fixed point.

   (c) Show that if f has exactly one fixed point, not equal to ∞, then $(\mathbb{C}_\infty, f)$ is conjugate to $(\mathbb{C}_\infty, g)$ where $g(z) = z + (2c)/(a + d)$.

   (d) Prove Theorem 11.6.

9. Theorem 11.9 characterizes the dynamics of Möbius transformations with two fixed points when $|k| \neq 1$. In this exercise, we consider several examples of Möbius transformations with two fixed points for which $|k| = 1$.

(a) Describe the fixed points and long term dynamics for the Möbius transformation $f(z) = -z$.

(b) Describe the fixed points and long term dynamics for the Möbius transformation $f(z) = i\,z$.

(c) Describe the fixed points and long term dynamics for the Möbius transformation with $a = \cos\left(\frac{2\pi}{n}\right) + i\sin\left(\frac{2\pi}{n}\right)$, $b = c = 0$, and $d = 1$:

$$f(z) = \left(\cos\left(\frac{2\pi}{n}\right) + i\sin\left(\frac{2\pi}{n}\right)\right) z.$$

(Hint: $\cos\left(\frac{2\pi}{n}\right) + i\sin\left(\frac{2\pi}{n}\right) = e^{\frac{2\pi}{n}i}$.)

(d) Suppose $f(z) = e^{i\pi^2} z$. Explicitly write out the iterates of various initial conditions under f and describe their long term behavior.

10. Figure 11.1 is a pictorial representation of a homeomorphism, called stereographic projection, between $\mathbb{C}$ and the unit sphere with the north pole removed. It sends a point P on the sphere to the point Q on $\mathbb{C}$ that intersects the line that goes through the point and the north pole, denoted by N. Construct the map thus described and prove that it is indeed a homeomorphism.

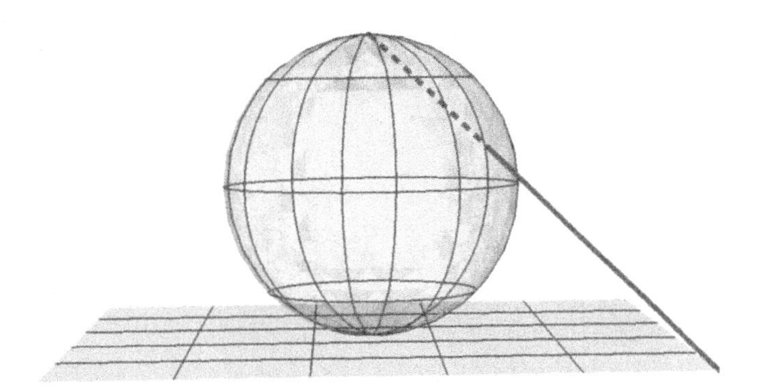

Figure 11.1. Stereographic projection from the unit sphere onto $\mathbb{C}$.

Project

Consider the dynamical system $(\mathbb{C}_\infty, f)$ where $f(z) = z^2$. Describe its dynamical properties, including all possible long term behaviors of its initial conditions. To what extent can the function f be generalized and still have these same dynamical properties?

Module 12

The Julia and Mandelbrot Sets

Exploration

Consider the dynamical system $(\mathbb{C}, f)$ where

$$f(z) = z^2 + 0.64\,i\,.$$

Can you describe the set of initial conditions with unbounded orbits for this dynamical system? With bounded orbits?

Exposition

Our investigation into the dynamics of complex-valued functions began in the last module where we studied properties of the systems $(\mathbb{C}_\infty, f)$ where f was a Möbius transformation. In this module we will study the dynamics of the family of systems $(\mathbb{C}, f_c)$ where $f_c(z) = z^2 + c$ and $c \in \mathbb{C}$. As with any family of functions, there are two approaches we might take when studying the dynamics of the family. First, we can study the dynamics of individual functions in the family by fixing the constant c and then varying the initial condition. This is the tactic used in the exploration for $c = 0.64i$. We could alternatively study the dynamics of the entire family by fixing a complex-valued initial condition z_0 and varying the constant c. This is the tactic that we took in Module 7 when we created bifurcation diagrams for several families of real-valued functions. Using these two approaches to study the family $(\mathbb{C}, f_c)$ leads to some of the most beautiful and iconic images in mathematics. Our goal in this module is to explore these images and their connections to familiar concepts from dynamical systems.

We begin with the first approach and fix the constant $c \in \mathbb{C}$, say $c = 0.29 + 0.54i$. By iterating a variety of initial conditions, it is not difficult to see that some initial conditions seem to have bounded orbits while others have unbounded orbits. Our focus here will be on the set of initial conditions with bounded orbits. When we investigated the set of initial conditions with bounded orbits for the dynamical system in Module 4, we discovered one of the most interesting sets in mathematics, the Cantor set Γ. As we will see, the set of initial conditions with bounded orbits for $(\mathbb{C}, f_c)$ is similarly interesting from a mathematical point of view. In addition, the set has remarkable geometric properties and is so visually striking that it has made its way into popular culture. We begin by giving a definition.

Definition 12.1. *Let $f_c(z) = z^2 + c$ for $c \in \mathbb{C}$. The set of all initial conditions with bounded orbits is denoted $\mathcal{F}_c$ and is called the* **filled-in Julia**[1] **set for** c. *The boundary of $\mathcal{F}_c$ is denoted $\mathcal{J}_c$ and is called the* **Julia set for** c.

In constructing Γ, we first noticed that if an iterate ever left the interval $[0, 1]$, its orbit would not be bounded. We then proceeded by finding Γ_1, the set of initial conditions that remained in $[0, 1]$ after one iteration, and then finding Γ_2, the set of initial conditions that remained in $[0, 1]$ after two iterations, and so on. At each stage, the set Γ_n of initial conditions that remain in $[0, 1]$ after n iterations gave us an approximate sense of Γ, and $\Gamma = \cap_{n \in \mathbb{N}} \Gamma_n$.

To find $\mathcal{F}_c$, we will proceed in a similar fashion and thus we need to first find a subset of $\mathbb{C}$ that plays the role of $[0, 1]$ in the construction of Γ. That is, we would like to identify a subset S of $\mathbb{C}$ with the property that if the iterates of an initial condition leave S, then that initial condition will not have a bounded orbit. The following theorem, which you will prove in Exercise 3, identifies such a subset for any f_c in our family.

Theorem 12.2. *Let $f_c(z) = z^2 + c$ where c is a complex-valued constant. If $|z_0| > max(|c|, 2)$, then $\lim_{n\to\infty} f_c^n(z_0) = \infty$.*

For $c = 0.29 + 0.54i$, we have $|c| \approx 0.6$ so $\max(|c|, 2) = 2$, and we set

$$S = \{z : |z| \leq 2\}.$$

[1] Gaston Julia (1893 - 1978) was only 25 when he published his 199 page paper *Mémoire sur l'iteration des fonctions rationelles*. He was very famous during the 1920s as a result of this work, but as time went on his contributions were forgotten. It took the invention of the computer to reignite interest in the iteration of functions, and Benoît Mandelbrot brought Julia's work back to prominence in the 1970s.

Table 12.1 contains the first five iterates of initial conditions z_1, z_2, and z_3. We see that $|z_1| \approx 3.6$ and so $z_1 \notin S$. Thus Theorem 12.2 tells us that the iterates of z_1 will grow without bound; the iterates shown in the table reflect this.

n	$f_c^n(z_1) =$ f_c^n(-2.25+2.75i)	$f_c^n(z_2) =$ f_c^n(0.75+0.25i)	$f_c^n(z_3) =$ f_c^n(-0.25+0.75i)
1	$-2.21 - 11.84i$	$0.79 + 0.92i$	$-0.21 + 0.17i$
2	$-134.9 + 52.90i$	$0.08 + 1.99i$	$0.31 + 0.47i$
3	$15403 - 142578i$	$-3.65 + 0.85i$	$0.16 + 0.83i$
4	$3.4 \times 10^{17} - 4.4 \times 10^8 i$	$12.88 - 5.63i$	$-0.37 + 0.81i$
5	$-1.9 \times 10^{17} - 3.0 \times 10^{16} i$	$134.5 - 144.i$	$-0.23 - 0.06i$

Table 12.1. Iterates of $f_{0.29+0.54i}$ for three initial conditions.

On the other hand, $|z_2| = |z_3| \approx 0.27$, and both z_2 and z_3 are in S. Theorem 12.2 does not tell us anything about the iterates of these initial conditions. However, we see in Table 12.1 that $|f_c^2(z_2)| > 2$; therefore by Theorem 12.2 the iterates of $f_c^2(z_2)$, and thus of z_2, will grow without bound. On the other hand, the iterates of z_3 displayed in Table 12.1 stay close to the origin, and the orbit of z_3 might be bounded.

The behavior of the iterates of z_2 shows us that not all initial conditions in S will have bounded orbits, because not all their iterates will stay in S. We would like to identify the subset of S consisting of initial conditions all of whose iterates stay in S. In Module 4, when creating Γ_n, we were able to determine the initial conditions in $[0, 1]$ that remained in $[0, 1]$ after n iterations by a careful study of the graph of the function f. We do not have a graphical representation for f_c, so we will develop a new technique based on iteration to approximate the initial conditions whose orbits remain in S.

If we could iterate every point in S and shade it black if the iterates appear to be bounded and leave it white otherwise, we would have a pretty good picture of $\mathcal{F}_c$. However, there are infinitely many initial conditions in S and we cannot iterate all of them. Instead, we will choose a finite number of sample initial conditions, iterate them to see if they remain in S, and assume that nearby initial conditions will behave similarly.

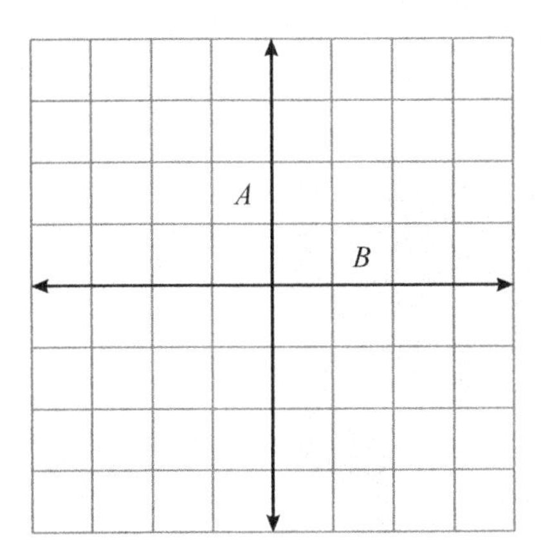

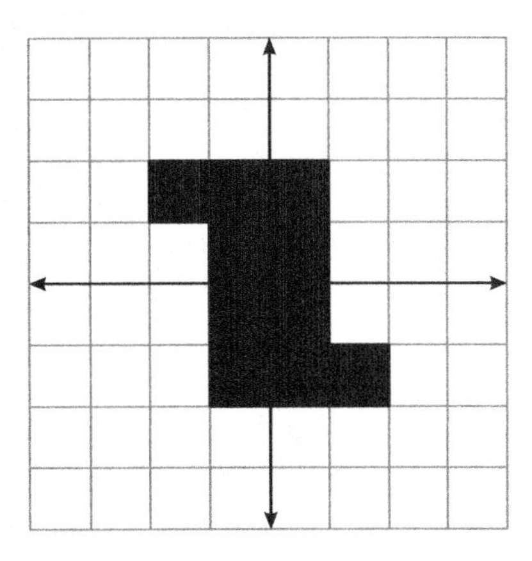

Figure 12.1. Approximating $\mathcal{F}_c$ using sixty-four sub-boxes and five iterations of f_c.

We make this more precise as follows: first notice that S is contained in the box $B = [-2, 2] \times [-2, 2]$. We will divide B into sixty-four smaller sub-boxes, each with side length 0.5, and we will use the middle point of each box as our sample initial condition. We will iterate f_c five times for those sample initial conditions and shade the entire sub-box according to the behavior of the sample initial condition. For example, the initial conditions $z_3 = -0.25 + 0.75i$ and $z_2 = 0.75 + 0.25i$ are the sample initial conditions that will determine the shading of the sub-boxes labeled A and B respectively in Figure 12.1. After five iterations, the iterates of z_3 appear to remain bounded so the sub-box A will be shaded black. After five iterations, the iterates of z_2 have left S, so the orbit of z_2 is unbounded and we will not shade the sub-box B.

We can further subdivide B, as shown in Figure 12.2, to get better and better approximations to $\mathcal{F}_c$, and when we do so, the intricacy and beauty of the filled-in Julia set $\mathcal{F}_c$ begins to reveal itself.

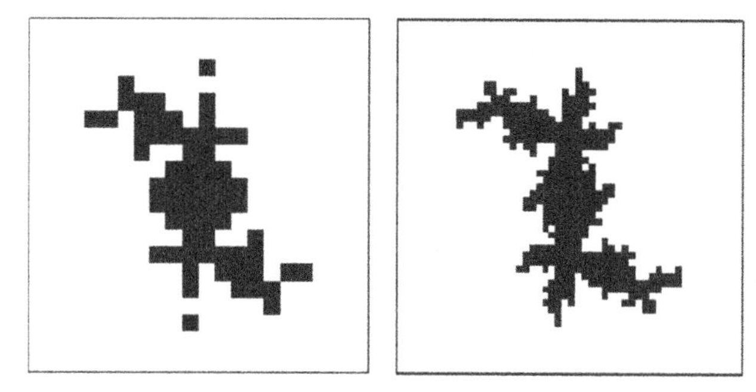
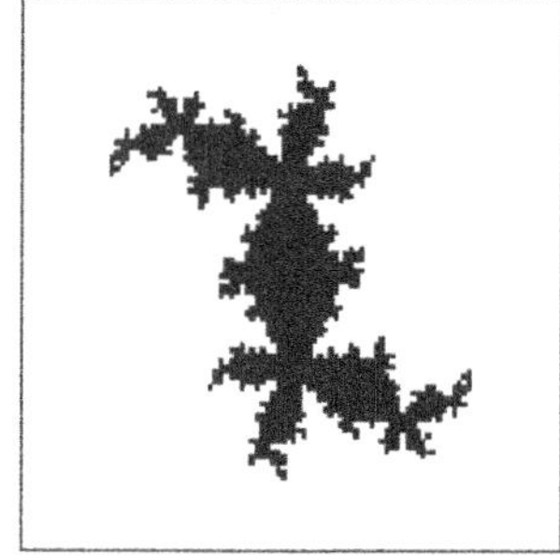

Figure 12.2. Using 20^2, 50^2, and 100^2 sub-boxes to approximate $\mathcal{F}_c$.

Like the set of initial conditions with bounded orbits Γ that we found in Module 4, the Julia set $\mathcal{J}_c$ seems to exhibit self-similarity. In Figure 12.3, we zoom in on the top left region of $\mathcal{F}_c$ occurring in $[-1.2, 0] \times [0.3, 1.1]$, and we see that the complexity of the boundary of the region is similar to the complexity of all of $\mathcal{J}_c$.

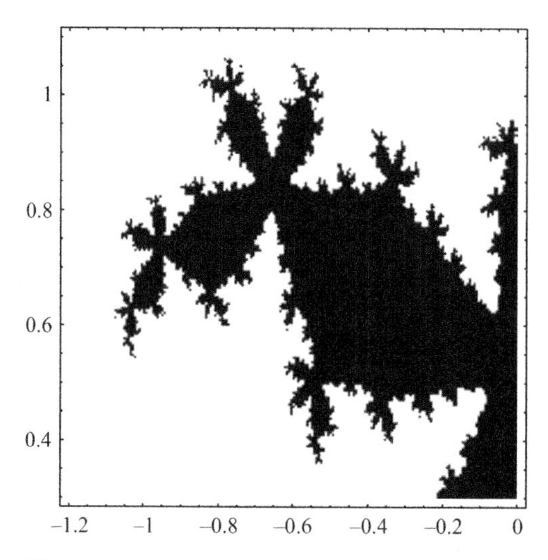

Figure 12.3. A close up view of one region in $\mathcal{F}_c$.

We can similarly study other functions in our family $(\mathbb{C}, f_c)$ by choosing other fixed values for c and looking for the filled-in Julia set $\mathcal{F}_c$. The collection of filled-in Julia sets, $\{\mathcal{F}_c\}_{c\in\mathbb{C}}$, leads to many rich and beautiful images; watch for them not only in your mathematics classes but also in popular culture decorating things from calendars to carpets. Figure 12.4 shows the filled-in Julia sets for three additional values of c.

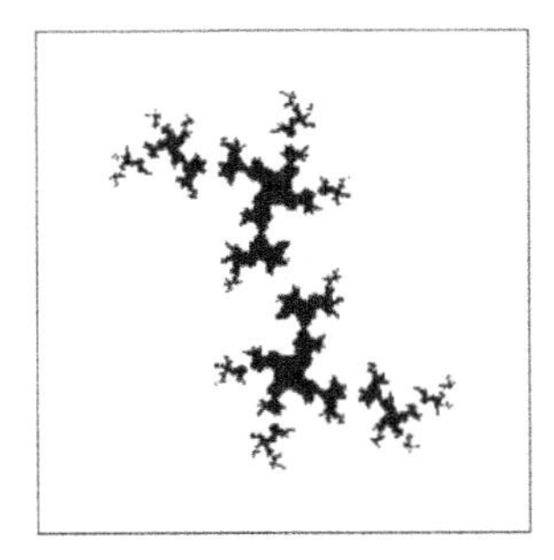
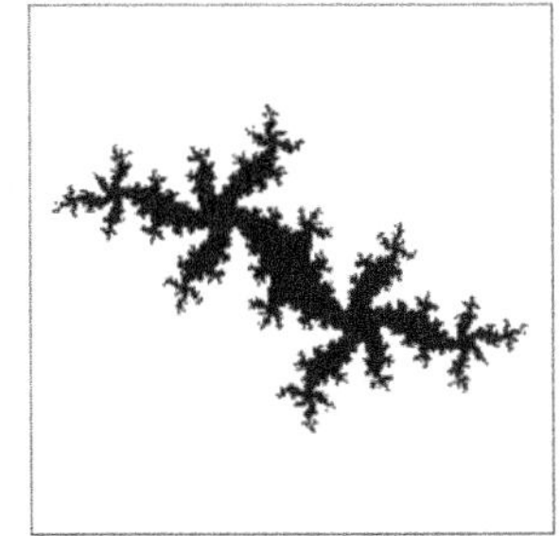
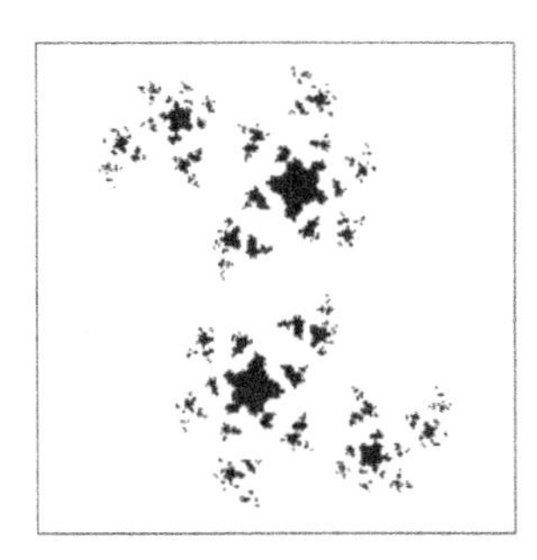

Figure 12.4. Filled-in Julia sets $\mathcal{F}_c$ for $c = 0.3 + 0.6i$, $-0.5 + 0.6i$, and $0.4 + 0.4i$.

We next turn our attention to the second approach described at the beginning of this module for studying the dynamics of the family $(\mathbb{C}, f_c)$. As we did in Module 7, we will now fix an initial condition, say $z_0 = 0 + 0i$, and we will vary the parameter c. We will iterate f_c at $z_0 = 0 + 0i$ for different c values and investigate any bifurcations, or changes in behavior of the iterates of z_0. Consistent with our work thus far on filled-in Julia sets, the change of behavior we will be interested in is whether or not the orbit of $z_0 = 0 + 0i$ is bounded.

	$c_1 = 0.3 + 0.6i$	$c_2 = -0.5 + 0.6i$	$c_3 = 0.4 + 0.4i$
$f_c^{16}(z_0)$	$-0.21 + 2.0i$	$-0.81 + 0.43i$	$\lvert f_c^{16}(z_0)\rvert > 10^8$
$f_c^{17}(z_0)$	$-3.5 - 0.21i$	$-0.02 - 0.09i$	$\lvert f_c^{17}(z_0)\rvert > 10^{17}$
$f_c^{18}(z_0)$	$12.4 + 2.1i$	$-0.51 + 0.60i$	$\lvert f_c^{18}(z_0)\rvert > 10^{35}$
$f_c^{19}(z_0)$	$151 + 52i$	$-0.61 - 0.01i$	$\lvert f_c^{19}(z_0)\rvert > 10^{71}$
$f_c^{20}(z_0)$	$20,101 + 15,657i$	$-0.13 + 0.62i$	$\lvert f_c^{20}(z_0)\rvert > 10^{142}$

Table 12.2. $f_c^n(0 + 0i)$ for three values of the parameter c.

For example, Table 12.2 shows the sixteenth through twentieth iterates of z_0 for three values of the parameter c. Since $|f_c^{20}(z_0)| > 2$ for c_1 and c_3, Theorem 12.2 tells us for these values of c, the orbit of z_0 is unbounded. For c_2, however, the orbit of z_0 could be bounded. If so, a bifurcation occurs for c somewhere between c_1 and c_2, and between c_3 and c_2, where the orbit of z_0 changes from being bounded to unbounded. As we did in Module 7, we would like to construct a bifurcation diagram to illustrate graphically the values of c for which these changes occur.

In Module 7, our parameter c was a real number, so when creating a bifurcation diagram, we could represent all values of c along the x-axis. For our current family of functions, $\{f_c\}_{c\in\mathbb{C}}$, our parameter c is a complex number so we will need the complex plane to represent the possible values of c. We can indicate when the orbit of z_0 is bounded for a particular c by shading that value of c dark. If the orbit of z_0 is unbounded for a particular c, we will not shade that value of c. Of course, there are infinitely many possible values of

c and we cannot test them all, so as we did in the construction of the filled-in Julia sets, we will divide the complex plane into sub-boxes, choose c in the center of each sub-box, and shade the entire sub-box according to the behavior of the iterates $f_c^n(z_0) = f_c^n(0 + 0i)$ with c given by the complex number in the center of the sub-box. This will give us an approximation of the bifurcation diagram we are looking for.

Theorem 12.2 again simplifies our work, because if $|c| > 2$, then

$$|f_c^2(z_0)| = |c^2 + c| = |c||c + 1| \geq |c|(|c| - 1) > |c|.$$

This tells us that $|f_c^2(z_0)| > \max(|c|, 2)$ and the orbit of $z_0 = 0 + 0i$ is unbounded. Thus the values of c for which the orbit of $z_0 = 0 + 0i$ will be bounded must lie in $B = [-2, 2] \times [-2, 2]$.

Figure 12.5 shows an approximation to our bifurcation diagram using sixty-four sub-boxes of B and, to determine shading, twenty iterations of f_c at $z_0 = 0 + 0i$ where c is the value in the center of each sub-box.

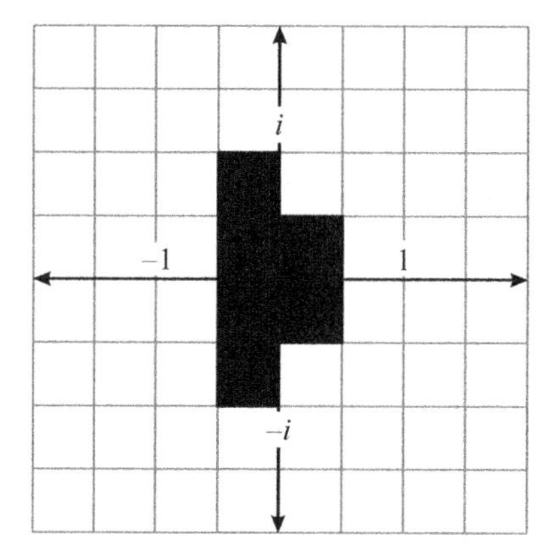

Figure 12.5. Sub-boxes for which iterates remain bounded are shaded.

In Figure 12.6, we again use twenty iterations of f_c at $z_0 = 0 + 0i$ where c is the value in the center of each sub-box for even more approximating sub-boxes. We shade the sub-box black if for all twenty iterations $|f_c^n(z_0)| \leq \max(|c|, 2)$, and we leave it unshaded otherwise.

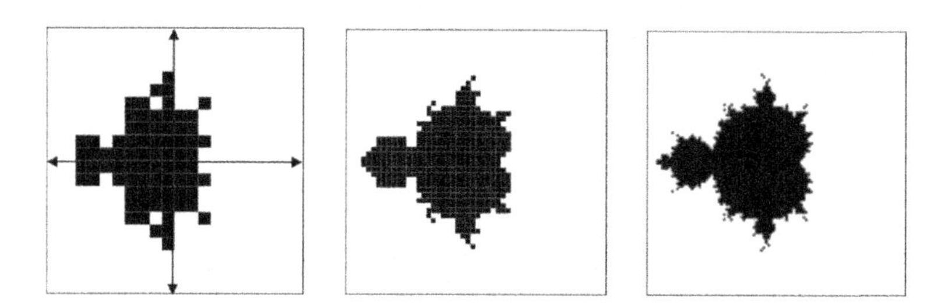

Figure 12.6. Using 20^2, 50^2, and 100^2 sub-boxes.

Our images in Figure 12.6 approximate a set known as the Mandelbrot set.

Definition 12.3. *The* **Mandelbrot**[2] **set** $\mathcal{M}$ *is defined by*

$$\mathcal{M} = \left\{ c \in \mathbb{C} : \{f_c^k(0+0i)\}_{k=0}^{\infty} \text{ is a bounded set} \right\}$$

where $f_c(z) = z^2 + c$.

The Mandelbrot set $\mathcal{M}$ is a type of bifurcation diagram in that the dynamical systems $(\mathbb{C}, f_c)$ with parameters $c \in \mathcal{M}$ exhibit very different long term behaviors (namely bounded orbits for $z_0 = 0 + 0i$) than the dynamical systems $(\mathbb{C}, f_c)$ with parameters $c \notin \mathcal{M}$. A bifurcation, or change in behavior, occurs for those parameters c on the boundary of the Mandelbrot set.

Like the Canter set Γ, the Julia and Mandelbrot sets are beautiful and iconic sets in mathematics that arise as a result of questions and techniques that are familiar to us. In the project, you will do further research to learn more about these fascinating sets, including the important relationship between them.

Exercises

1. (a) Consider the real-valued quadratic function $f(x) = x^2 - 0.75$ where $x \in \mathbb{R}$. Find and classify the real-valued fixed points of f. For any attracting fixed point, find the corresponding basin of attraction. Describe the long-term behavior of real-valued initial conditions not in any basin of attraction.

   (b) Now let $f_{-0.75}(z) = z^2 - 0.75$ be the corresponding complex-valued quadratic function. Discuss the long term behavior of purely imaginary initial conditions $z = ai, a \in \mathbb{R}$.

   (c) Use the first two parts of this exercise to describe the filled-in Julia set $\mathcal{F}_{-0.75}$ on the real and imaginary axes.

   (d) Use a Julia set generator from the internet, or program your own, to produce an approximate image of $\mathcal{F}_{-0.75}$ and verify that the image that is created is consistent with your answer to (c).

2. (a) Consider the real-valued quadratic function $f(x) = x^2 - 1$ where $x \in \mathbb{R}$. Show that f has no attracting fixed points, but that it does have an attracting 2-cycle. Find the basin of attraction for the 2-cycle, and describe what happens to the iterates of real-valued initial conditions not in it.

   (b) Now let $f_{-1}(z) = z^2 - 1$ be the corresponding complex-valued quadratic function. Discuss the long term behavior of purely imaginary initial conditions $z = ai, a \in \mathbb{R}$.

   (c) Use the first two parts of this exercise to describe the filled-in Julia set $\mathcal{F}_{-1}$ on the real and imaginary axes.

---

[2]Benoît Mandelbrot (1924–2010) was born in Poland and emigrated to France when he was twelve years old. He spent most of his career at the IBM Watson Research Center and became the Sterling Professor of Mathematical Sciences at Yale University after his retirement. With the accessibility of computing power at IBM, Mandelbrot became interested in Julia's 1918 paper and showed it was the source of some of the most beautiful fractals known.

(d) Use a Julia set generator from the internet, or program your own, to produce an approximate image of $\mathcal{F}_{-1}$ and verify that the image that is created is consistent with your answer to (c).

3. (a) Let $f_c(z) = z^2 + c$, $c \in \mathbb{C}$, and suppose that $|z_0| > \max\{|c|, 2\}$. Show that $|f(z_0)| \geq \lambda |z_0|$ for some $\lambda > 1$.
   (b) Prove Theorem 12.2.
   (c) Show that if there exists $n \in \mathbb{N}$ with $|f^n(z_0)| > \max\{|c|, 2\}$, then $z_0 \notin \mathcal{F}_c$.

4. In this exercise, we explore some properties of $\mathcal{F}_c$.
   (a) Show that $\mathcal{F}_c$ is bounded for all $c \in \mathbb{C}$. That is, show that for each $c \in \mathbb{C}$ there exists $R > 0$ such that if $z \in \mathcal{F}_c$, then $|z| \leq R$.
   (b) Show that $\mathcal{F}_c$ is symmetric for all $c \in \mathbb{C}$. That is, show that if $z \in \mathcal{F}_c$, then $-z \in \mathcal{F}_c$ as well. Verify that the images in Figures 12.2 and 12.4 are symmetric.
   (c) Show that $\mathcal{F}_c$ is an invariant set for all $c \in \mathbb{C}$. That is, show that $f_c(\mathcal{F}_c) = \mathcal{F}_c$.
   (d) Show that there exists $\epsilon > 0$ such that for every $z \notin \mathcal{F}_c$, there is $k \in \mathbb{N}$ with the distance between $f_c^k(z)$ and $\mathcal{F}_c$ larger than ϵ.

5. (a) Use a Julia set generator from the internet, or program your own, to create approximate images of the filled-in Julia set $\mathcal{F}_c$ for a variety of values $c \in \mathbb{C}$. Choose some values of c in the Mandelbrot set $\mathcal{M}$ and some values of c outside $\mathcal{M}$.
   (b) Based on the images that you created in (a), can you conjecture a relationship between the properties of $\mathcal{F}_c$ and whether or not $c \in \mathcal{M}$?
   (c) In Exercises 4a and 4b, you proved that $\mathcal{F}_c$ is bounded and symmetric for all $c \in \mathbb{C}$. Verify that the images you created in (a) are bounded and symmetric.

6. Many of the filled-in Julia set generators found on the internet produce pictures that are quite colorful. Choose a web generator that produces colorful images of filled-in Julia sets and determine the mathematical meaning of the colors.

7. In this question, we study the dynamics of a general complex-valued quadratic function $p(z) = az^2 + bz + d$ where $a, b, d \in \mathbb{C}$ and $a \neq 0 + 0i$.
   (a) Let $\phi(z) = \frac{2z-b}{2a}$. Show that $\phi : \mathbb{C} \to \mathbb{C}$ is a homeomorphism and find ϕ^{-1}.
   (b) Find $c \in \mathbb{C}$ such that ϕ is a conjugacy between $(\mathbb{C}, f_c)$ and $(\mathbb{C}, p)$.
   (c) Let $p(z) = z^2 + 4z + 2$. Find ϕ, ϕ^{-1}, and c as described in (a) and (b). Describe the filled-in Julia set for the resulting $f_c(z)$ and how it relates to the set of bounded orbits for $p(z)$.

Project

This module only scratches the surface of the many interesting things known about the filled-in Julia sets $\{\mathcal{F}_c\}_{c \in \mathbb{C}}$ and the Mandelbrot set $\mathcal{M}$. In particular, we have only hinted at the fascinating relationship between $\{\mathcal{F}_c\}_{c \in \mathbb{C}}$ and $\mathcal{M}$ (Exercise 5b). There are many books and reputable internet resources on the subject. For this project, do further research with the goal of understanding as much as possible about the relationship between $\{\mathcal{F}_c\}_{c \in \mathbb{C}}$ and $\mathcal{M}$.

Module 13

Symbolic Dynamical Systems

Exploration

Let X consist of all possible doubly-infinite sequences

$$x = \dots x_{-2}x_{-1} \,.\, x_0x_1x_2x_3 \dots$$

where each term x_i of the sequence is either the symbol 0 or the symbol 1. Define the **shift map** $\sigma : X \to X$ as follows:

$$\sigma(x) = \sigma(\dots x_{-2}x_{-1} \,.\, x_0x_1x_2x_3 \dots) = \dots x_{-2}x_{-1}x_0 \,.\, x_1x_2x_3x_4 \dots .$$

Compute $\sigma^n(x)$ for a variety of values of n and a variety of sequences from X. What are all the periodic points of the dynamical system (X, σ)?

Now let X_1 be the subset of X consisting of all possible doubly-infinite sequences of 0s and 1s subject to the constraint that there is an even number of 0s between any two occurrences of the symbol 1. Answer the same question for the dynamical system (X_1, σ).

Let X_2 be the subset of X consisting of all possible doubly-infinite sequences of 0s and 1s subject to the constraint that a 1 may never follow a 0. Answer the same question for the dynamical system (X_2, σ).

Find a subset X_3 of X such that (X_3, σ) has a different set of periodic points.

Exposition

We are all familiar with the experience of turning on a computer, retrieving files, reading or editing them, and then saving them on a hard drive or on some other medium. Your computer stores data as a sequence of electronic impulses and retrieves the data by using a device designed to detect these impulses. The impulses are either there or not, and the device simply gives that information: impulse on, impulse off, impulse off, etc. For convenience, we will denote "impulse on" as the symbol 1 and "impulse off" as the symbol 0. Thus data stored on a computer can be viewed as a string of 0s and 1s. For instance, impulse on, impulse off, impulse off becomes the string 100.

A string of 0s and 1s representing actual data is potentially very long, so we imagine all doubly-infinite sequences of 0s and 1s as representing all possible data. The set of all possible doubly-infinite sequences of 0s and 1s, namely the set X encountered in the exploration, is denoted by $\{0, 1\}^{\mathbb{Z}}$ where $\mathbb{Z}$ represents the set of integers. We can think of a computer reading the data encoded in a sequence from $\{0, 1\}^{\mathbb{Z}}$ as currently reading the symbol immediately to the right of the period and then shifting that symbol over to the left of the period. It then reads the new symbol to the right of the period, and so on. This process is modeled by the shift map σ encountered in the exploration. Together the phase space $\{0, 1\}^{\mathbb{Z}}$ and the shift map σ define a dynamical system $(\{0, 1\}^{\mathbb{Z}}, \sigma)$ called the **full shift on two symbols**.

The set $\{0, 1\}$ is called the **alphabet** of the dynamical system $(\{0, 1\}^{\mathbb{Z}}, \sigma)$. In general, if $\mathcal{A} = \{0, 1, 2, \ldots, n-1\}$, then $\mathcal{A}^{\mathbb{Z}}$ represents the set of all doubly-infinite sequences consisting of symbols from the set $\mathcal{A}$. The resulting dynamical system, $(\mathcal{A}^{\mathbb{Z}}, \sigma)$, is called the **full shift on n symbols**.

All the dynamical systems introduced in the exploration have phase spaces that consist of doubly-infinite sequences constructed using the alphabet $\mathcal{A} = \{0, 1\}$; however, only the first example is the full shift. The phase spaces of the other examples are subsets of $\{0, 1\}^{\mathbb{Z}}$ obtained by eliminating from $\{0, 1\}^{\mathbb{Z}}$ all sequences that do not obey a certain rule. This type of restriction on the phase space is often necessary for applications. To illustrate this, let us return to the example of computer data storage.

Data are stored on magnetic storage devices consisting of a sequence of bar magnets. The electrical impulse indicating the symbol 1 is generated by a change in polarity from one magnet to the next. The string 1001100 would be stored as illustrated in Figure 13.1.

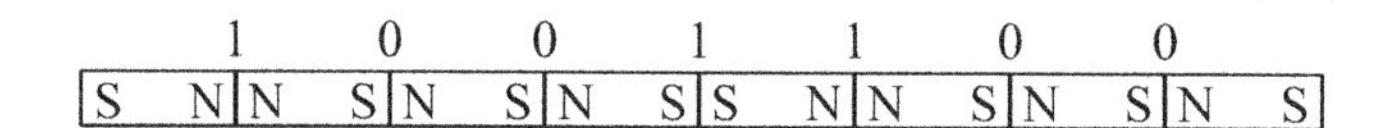

Figure 13.1. Storing 1001100 in a magnetic medium.

Depending on the hardware used, storing the string illustrated in Figure 13.1 could lead to a complication known as intersymbol interference. Informally, this can be described by saying that if polarity changes occur too close together, they cancel each other out. Another hardware complication known as clock drift prevents us from having a long run of 0s. This is because, when reading a magnetic medium, the computer detects a 0 by a period of time in which there is no electrical impulse. Several 0s in a row are detected by a long period of time with no electrical impulse. When the time elapsed without an electrical impulse is too long, it may be difficult to know exactly how many 0s the time period represents.

A common method used to deal with these hardware limitations is to restrict the allowable sequences to those for which there are no consecutive 1s and for which there is a restriction on the number of consecutive 0s. The first restriction will prevent intersymbol interference and the second will prevent clock drift. If, for example, X represents the set of sequences with no consecutive 1s and at most three consecutive 0s, then (X, σ) is called the **(1, 3)-run length limited shift**. Other variations are also commonly used; if we allow only the set of sequences for which there are at least a and no more than b occurrences of 0s between any two 1s, the resulting set of sequences is the phase space of the **(a, b)-run length limited shift**.

The set of sequences in the phase space of the $(1, 3)$-run length limited shift consists exactly of the sequences from $\{0, 1\}^{\mathbb{Z}}$ for which the strings $\{11, 0000\}$ never occur. This idea of forbidding certain patterns from appearing in a sequence is an important one, and we will return to it after we introduce some necessary terminology.

We define a **block** (or **word**) over an alphabet $\mathcal{A}$ to be a finite sequence of symbols from $\mathcal{A}$. The **length of a block** is the number of symbols it contains. A **k-block** will refer to a block of length k. We say a block u **occurs in a sequence** $x \in \mathcal{A}^{\mathbb{Z}}$ if there are indices $i \leq j$ such that x restricted to $x_i x_{i+1} \ldots x_j = x_{[i,j]} = u$. A block u is said to **occur in a set X** if there is a sequence x in X such that u occurs in x. We can use this new terminology to describe the phase space of the $(1, 3)$-run length limited shift as the set of sequences from $\{0, 1\}^{\mathbb{Z}}$ in which the blocks $\mathcal{F} = \{11, 0000\}$ do not occur.

Definition 13.1. *Let $\mathcal{A}$ be a finite alphabet and let $\mathcal{F}$ be a collection of blocks over $\mathcal{A}$. The* **shift space X given by** *$\mathcal{F}$ is the subset of $\mathcal{A}^{\mathbb{Z}}$ given by*

$$X = \{x \in \mathcal{A}^{\mathbb{Z}} : \text{ no block from } \mathcal{F} \text{ occurs in } x\}.$$

The phase space of the $(1, 3)$-run length limited shift and the spaces X_1 and X_2 encountered in the exploration are examples of shift spaces (Exercise 1). The set $\{0, 1\}^{\mathbb{Z}}$ is also a shift space with $\mathcal{F} = \emptyset$.

It is not hard to see that a shift space X is **shift invariant**, meaning that if x is a sequence in X, then $\sigma(x)$ is also in X.

Definition 13.2. *A* **symbolic dynamical system** *is a pair (X, σ) where the phase space X is a shift space given by a collection $\mathcal{F}$ of forbidden blocks and the function σ is the shift map.*

All symbolic dynamical systems involve the function $\sigma : X \to X$ but, as suggested in the exploration, by varying the shift space X we will see a variety of different dynamical behaviors. Because of this we often give the same name to the dynamical system (X, σ) and the shift space X. For example, both $(\{0, 1\}^{\mathbb{Z}}, \sigma)$ and $\{0, 1\}^{\mathbb{Z}}$ are often called the full shift.

One of the behaviors of interest thus far has been the existence and nature of fixed points, but Theorems 1.3 and 1.4 which classify fixed points as either attracting or repelling depend on the notion of distance in $\mathbb{R}$. In order to discuss these concepts for symbolic dynamical systems we need a notion of distance between sequences in a shift space. We will define this distance in such a way that two sequences x and y in a shift space X are close if they agree on a long block centered at the zeroth position. Formally the distance function ρ is defined as:

$$\rho(x, y) = \begin{cases} 2 & \text{if } x_0 \neq y_0 \\ 2^{-k} & \text{if } x \neq y \text{ and } k \text{ is the largest number such that } x_{[-k,k]} = y_{[-k,k]} \\ 0 & \text{if } x = y. \end{cases}$$

For example, let

$$x = \ldots 0\,0\,0.0\,1\,1\,0\,0\,0 \ldots$$
$$y = \ldots 0\,0\,0.0\,0\,1\,0\,0\,0 \ldots$$
$$z = \ldots 1\,0\,0.0\,1\,1\,0\,0\,1 \ldots.$$

Since x and y agree at the symbol in the zeroth position but disagree at the symbol in the first position, $\rho(x, y) = 2^0 = 1$. On the other hand, the largest block centered at the zeroth position on which x and z agree is on the symbols at indices from -2 to 2, since they disagree on the symbol at position -3. Thus $\rho(x, z) = 2^{-2} = 1/4$.

The distance between two points in a shift space X is at most 2, and thus all orbits are bounded. In this way, symbolic dynamical systems are similar to the system studied in Module 4.

Using ρ to determine the distance between two doubly-infinite sequences of symbols seems quite different from using the absolute value function for points in $\mathbb{R}$ or the Euclidean metric on $\mathbb{R}^2$ or $\mathbb{C}$. However, these notions of distance share many of the same properties (Exercise 4). Thus our previous definitions that made use of the notion of distance on $\mathbb{R}$ can now apply to symbolic dynamical systems.

For instance, suppose (X, σ) is a symbolic dynamical system and $p \in X$ is a fixed point. In the exploration you will have noticed that fixed points consist of a single repeated symbol a:

$$p = \quad \ldots.aaaa.aaaaaa\ldots. \ .$$

Suppose that p is attracting. Then Definition 1.3 tells us that there exists $\epsilon > 0$ such that every $x \in X$ with $\rho(x, p) < \epsilon$ satisfies $\lim_{n\to\infty} \sigma^n(x) = p$. Let us consider what this definition means in the new context of symbolic dynamical systems.

First, what does it mean to say $\rho(x, p) < \epsilon$? By the definition of ρ, $\rho(x, p) < \epsilon$ means the sequences x and p agree on positions $-k$ to k for some $k \in \mathbb{N}$ with $2^{-k} < \epsilon$. Thus $x_i = a$ for $-k \leq i \leq k$.

Next, what does it mean to say that $\lim_{n\to\infty} \sigma^n(x) = p$? Definition 1.2 tells us that $\lim_{n\to\infty} \sigma^n(x) = p$ implies that given $\tilde{\epsilon}$, there exists $N \in \mathbb{N}$ such that for all $n > N$, $\rho(\sigma^n(x), p) < \tilde{\epsilon}$. We use $\tilde{\epsilon}$ in this definition so as not to confuse it with the ϵ given to us by the fact that p is attracting. But $\tilde{\epsilon}$ can equal any value, so choose $\tilde{\epsilon} = 1/2$ and find N such that for all $n > N$, $\rho(\sigma^n(x), p) < \tilde{\epsilon} = 1/2$. Since $N + 1 > N$, as discussed in the previous paragraph this means that $\sigma^{N+1}(x)_0 = a$. But by the definition of σ, $\sigma^{N+1}(x)_0 = x_{N+1}$ and so we have $x_{N+1} = a$. Then, $N + 2 > N$, so $\sigma^{N+2}(x)_0 = x_{N+2} = a$. Continuing in this way, we see that $\sigma^n(x)_0 = x_n = a$ for all $n > N$.

Putting this all together, p is an attracting fixed point of a symbolic dynamical system if there exists $k \in \mathbb{N}$ such that all points x with $x_i = a$ for $-k \leq i \leq k$ also have $x_i = a$ for all large enough i. In Exercises 5c and 7, you will continue the work of understanding definitions we have used in this textbook in terms of the structure of the points in a shift space.

We conclude this module by considering an important variant of Definition 13.2. The shift map in this definition, $\sigma : X \to X$ where $X \subset \mathcal{A}^{\mathbb{Z}}$, is always an invertible map, as were the Möbius transformations of Module 11 and many of the linear maps of Module 10. This is in contrast to the noninvertible logistic family of maps $g_c : \mathbb{R} \to \mathbb{R}$ introduced in Module 1. We can define a new symbolic system that is noninvertible by simply changing the phase space to consist of one-sided sequences. More precisely, for any alphabet $\mathcal{A}$

define the set $\mathcal{A}^{\mathbb{N}}$ to be all one-sided infinite sequences consisting of symbols from $\mathcal{A}$:

$$\mathcal{A}^{\mathbb{N}} = \{.x_0x_1x_2\ldots \text{ where } x_i \in \mathcal{A} \text{ for each } i \geq 0\}.$$

Given a collection of blocks $\mathcal{F}$, we can define a shift space $X \subset \mathcal{A}^{\mathbb{N}}$ analogously to Definition 13.1 and the shift map σ on $x = .x_0x_1x_2.... \in X$ by

$$\sigma(x) = .x_1x_2x_3\ldots .$$

(X, σ) is then called a **one-sided symbolic dynamical system**. Adapting the distance function ρ to these symbolic systems (Exercise 3), we can extend the definition of attracting and repelling fixed points in this situation.

The new and abstract ideas in symbolic dynamical systems can feel foreign at first. The exercises in this module will develop intuition and skill in this new context. In the next module we will see that in addition to being interesting in their own right, symbolic dynamical systems are also a valuable tool in the study of all dynamical systems. In particular, we will investigate conjugacies between symbolic systems and some previously studied examples.

Exercises

1. Find a set $\mathcal{F}$ of forbidden blocks that gives rise to each of the following shift spaces. Describe the blocks in $\mathcal{F}$ as succinctly as possible. Give several examples of points from each shift space.

   (a) The shift space $X_1 \subseteq \{0, 1\}^{\mathbb{Z}}$ described in the exploration. X_1 is called the **even shift**.

   (b) The shift space $X_2 \subseteq \{0, 1\}^{\mathbb{Z}}$ described in the exploration.

   (c) The shift space $X_3 \subseteq \{0, 1\}^{\mathbb{Z}}$ you constructed in the exploration.

   (d) The shift space $Z \subseteq \{0, 1\}^{\mathbb{Z}}$ of all sequences where two consecutive 1s do not occur. Z is called the **golden mean shift**.

   (e) The set $W \subseteq \{a, b, c\}^{\mathbb{Z}}$ with the property that a block of the form ab^mc^ka may occur only if $m = k$, where b^m denotes the m-block consisting of m consecutive bs and c^k denotes the k-block consisting of k consecutive cs. W is called the **context free shift**.

   (f) The set $Y \subseteq \{-1, 1\}^{\mathbb{Z}}$ with the property that the sum of the symbols $+1$ and -1 occurring in every block is less than three in absolute value. Y is called the **charge constrained shift with charge two**.

2. (a) Find the distance between the given pairs of sequences.

   i. $x = \ldots 1100100.0010010011\ldots$ and $y = \ldots 1100100.0010011111\ldots$

   ii. $x = \ldots 0000001.0011000000\ldots$ and $y = \ldots 0000000.0010000000\ldots$

   (b) Find two points x, y from $\{0, 1\}^{\mathbb{Z}}$ for which $\rho(x, y) = 1/4$. Can you find a third point z so that $\rho(x, z)$ and $\rho(y, z)$ are also $1/4$?

   (c) Find examples of $x, y \in \{0, 1\}^{\mathbb{Z}}$, $x \neq y$, with the property that $\rho(x, y) < 1/10$.

3. (a) Let x and y be two one-sided sequences. Modify the definition of the distance function ρ from the exposition to define $\rho(x, y)$.

   (b) Use your definition to find the distance between the given pairs of one-sided sequences.

      i. $x = .01100111100\ldots$ and $y = .01100101100\ldots$

      ii. $x = .0000110000\ldots$ and $y = .0000100000\ldots$

   (c) Find examples of $x, y \in \{0, 1\}^{\mathbb{N}}, x \neq y$, with the property that $\rho(x, y) < 1/5$.

4. A **metric**, or distance function, on a set X is a function $\rho : X \times X \to \mathbb{R}$ such that for all $x, y, z \in X$

   i) $\rho(x, y) \geq 0$ with $\rho(x, y) = 0$ if and only if $x = y$

   ii) $\rho(x, y) = \rho(y, x)$

   iii) $\rho(x, z) \leq \rho(x, y) + \rho(y, z)$.

   The set X together with a metric on it is called a **metric space**.

   (a) Show that the usual distance function $d(x, y) = |x - y|$ is a metric on $\mathbb{R}$.

   (b) Show that the function ρ defined in the exposition is a metric on any shift space $X \subset \mathcal{A}^{\mathbb{Z}}$.

   (c) Does your answer for (b) change if $X \subset \mathcal{A}^{\mathbb{N}}$ and you use the distance function you defined in Exercise 3a?

5. Consider a sequence of points from $\{0, 1\}^{\mathbb{Z}}$. As each point in this sequence is itself a sequence, notating such a sequence is a bit tricky: let $x^{(n)}$ denote the nth element of the sequence, and as usual let x_m denote the mth position of the point x. For example, $x_m^{(n)}$ is the symbol in the mth position of the nth point in the sequence.

   As an example, for each $n \in \mathbb{N}$, define a point $x^{(n)}$ in $\{0, 1\}^{\mathbb{Z}}$ by setting $x_i^{(n)} = 0$ for all $i \neq n$ and $x_n^{(n)} = 1$.

   (a) Find $x_1^{(1)}, x_2^{(1)}, x_3^{(2)}$, and $x_4^{(2)}$.

   (b) Find $x^{(1)}, x^{(2)}, x^{(3)}$, and $x^{(4)}$.

   (c) Suppose a sequence of points $\{y^{(n)}\}_{n \in \mathbb{N}}$ in $\{0, 1\}^{\mathbb{Z}}$ converges to a point y in $\{0, 1\}^{\mathbb{Z}}$. Describe what this means in terms of the structure of the points $y^{(n)}$.

   (d) Consider the sequence of points $\{x^{(n)}\}_{n \in \mathbb{N}}$ described above. Does it converge in $\{0, 1\}^{\mathbb{Z}}$? If so, identify the limit and prove your claim.

6. Let $y = \ldots 0101.0101\ldots$ be a point in $\{0, 1\}^{\mathbb{Z}}$. Find a sequence of points $\{x^{(n)}\}_{n \in \mathbb{N}}$ in $\{0, 1\}^{\mathbb{Z}}$ so that $\rho(x^{(n)}, y) = 2^{-n}$, and show that $\{x^{(n)}\}_{n \in \mathbb{N}}$ converges to y.

7. Similar to what was done in the exposition for an attracting fixed point, explain the definition of a repelling fixed point in the context of a symbolic dynamical system.

8. (a) Is the fixed point $p = \ldots 111.1111\ldots$ in $(\{0, 1\}^{\mathbb{Z}}, \sigma)$ repelling?

   (b) Is the fixed point $p = \ldots 111.1111\ldots$ in the even shift attracting?

   (c) Give an example of a two-sided symbolic dynamical system with attracting and repelling fixed points.

9. (a) Is the fixed point $p = .1111\ldots$ in $(\{0,1\}^{\mathbb{N}}, \sigma)$ repelling?
   (b) Define the one-sided even shift using the same collection $\mathcal{F}$ as used in Exercise 1a. Is the fixed point $p = .1111\ldots$ in this system attracting?
   (c) Show that all the periodic points in $(\{0,1\}^{\mathbb{N}}, \sigma)$ are repelling.
10. Give an example of a subset X of $\{0,1\}^{\mathbb{Z}}$ that is not shift invariant. What would be the problem in using it as the phase space of a dynamical system?
11. Show that given a point $x \in \{0,1\}^{\mathbb{Z}}$ and $\epsilon > 0$ there is a periodic point $y \in \{0,1\}^{\mathbb{Z}}$ such that $\rho(x, y) < \epsilon$. This proves that periodic points are dense in $\{0,1\}^{\mathbb{Z}}$.

Project

In this module, we considered shift spaces where the points consisted of doubly infinite strings of symbols. Now let us consider a shift space that consists of two-dimensional arrays of symbols. The space of all such arrays of 0s and 1s is indicated by $\{0,1\}^{\mathbb{Z}^2}$. An example of a point in $\{0,1\}^{\mathbb{Z}^2}$ is shown in Figure 13.2. The location of a symbol is now described by an element of $\mathbb{Z}^2$; we indicate the symbol at the $(0,0)$th position by underlining. The configuration in Figure 13.2 also shows three symbols in bold. To familiarize yourself with the coordinate system, identify the two that are in locations $(-3,2)$ and $(2,-1)$ and give the location of the third bold symbol.

$$\begin{array}{ccccccccc}
 & & & & \vdots & & & & \\
 & \mathbf{1} & 1 & 1 & 0 & 0 & 0 & 0 & \\
 & 0 & 0 & 0 & 1 & \mathbf{0} & 1 & 0 & \\
\cdots & 1 & 1 & 0 & \underline{0} & 0 & 0 & 0 & \cdots \\
 & 0 & 0 & 1 & 1 & 1 & \mathbf{1} & 0 & \\
 & 1 & 0 & 0 & 0 & 0 & 0 & 0 & \\
 & & & & \vdots & & & &
\end{array}$$

Figure 13.2. A point in $\{0,1\}^{\mathbb{Z}^2}$.

Given $(m,n) \in \mathbb{Z}^2$, we can define a new shift map $\sigma^{(m,n)}$ that moves the symbols of a point to the left m units and down n units. An example of a point x and its shift $y = \sigma^{(2,1)}(x)$ is shown in Figure 13.3. The bold symbol in the point y shows the location of the symbol that was in position $(0,0)$ prior to the shift.

$$\begin{array}{ccccccccc}
 & & & & \vdots & & & & \\
 & 1 & 1 & 1 & 0 & 0 & 0 & 0 & \\
 & 0 & 0 & 0 & 1 & 0 & 1 & 0 & \\
\cdots & 1 & 1 & 0 & \underline{0} & 0 & 0 & 0 & \cdots \\
 & 0 & 0 & 1 & 1 & 1 & 1 & 0 & \\
 & 1 & 0 & 0 & 0 & 0 & 0 & 0 & \\
 & & & & \vdots & & & & \\
 & & & & x & & & &
\end{array}
\qquad
\begin{array}{ccccccccc}
 & & & & \vdots & & & & \\
 & * & * & * & * & * & * & * & \\
 & 1 & 0 & 0 & 0 & 0 & * & * & \\
\cdots & 0 & 1 & 0 & \underline{1} & 0 & * & * & \cdots \\
 & 0 & \mathbf{0} & 0 & 0 & 0 & * & * & \\
 & 1 & 1 & 1 & 1 & 0 & * & * & \\
 & & & & \vdots & & & & \\
 & & & & y & & & &
\end{array}$$

Figure 13.3. A point $x \in \{0,1\}^{\mathbb{Z}^2}$ and its shift $y = \sigma^{(2,1)}(x)$.

The dynamical system $(\{0,1\}^{\mathbb{Z}^2}, \sigma)$ where σ denotes the entire family $\{\sigma^{(m,n)}\}_{(m,n)\in\mathbb{Z}^2}$ is called the **two-dimensional full shift**.

Investigate what it means for a point to be periodic for this dynamical system. Give examples of periodic points that illustrate new behavior that results from the presence of an additional dimension.

$$
\begin{array}{cccccccccccc}
 & & & & & \vdots & & & & & & \\
 & 0 & 0 & 0 & 0 & 0 & 0 & 0 & 0 & 0 & 0 & \\
 & 1 & 0 & 0 & 1 & 0 & 0 & 1 & 0 & 0 & 1 & \\
 & 0 & 0 & 0 & 0 & 0 & 0 & 0 & 0 & 0 & 0 & \\
\cdots & 0 & 0 & 0 & 0 & \underline{0} & 0 & 0 & 0 & 0 & 0 & \cdots \\
 & 0 & 0 & 1 & 0 & 0 & 1 & 0 & 0 & 1 & 0 & \\
 & 0 & 0 & 0 & 0 & 0 & 0 & 0 & 0 & 0 & 0 & \\
 & & & & & \vdots & & & & & &
\end{array}
$$

Figure 13.4. A point in X.

Now let X be the subset of points from $\{0,1\}^{\mathbb{Z}^2}$ for which each 3×3 subarray of symbols contains exactly one 1. Thus the point in Figure 13.2 is not in X. The point shown in Figure 13.4, at least as far as we can tell from the portion of the point shown, could be in X. Find a collection of two-dimensional forbidden blocks $\mathcal{F}$ that gives rise to X. Investigate the set of periodic points for the dynamical system (X, σ).

This example can be generalized to systems that include only points with exactly two 1s in each 3×3 subarray. Investigate the set of periodic points for this generalization.

This project is based on [E] where various examples of two-dimensional symbolic dynamical systems are studied.

Module 14

Symbolic Dynamical Systems and Conjugacy

Exploration

Consider the function $f : \{0, 1\}^{\mathbb{Z}} \to \{0, 1\}^{\mathbb{Z}}$ given by

$$f(x)_i = \begin{cases} 1 & \text{if } x_i = 0 \\ 0 & \text{if } x_i = 1. \end{cases}$$

Is f a one-to-one function? Is it onto? What would it mean for f to be continuous?

Answer the same questions for the function $g : \{0, 1\}^{\mathbb{N}} \to \mathbb{R}$ given by

$$g(x) = \begin{cases} \displaystyle\sum_{i=0}^{\infty} x_i & \text{if the series is convergent} \\ 0 & \text{otherwise.} \end{cases}$$

Exposition

The symbolic dynamical systems introduced in the last module might at first seem exotic, but some of them exhibit dynamical behavior similar to other systems that we have studied. The one-sided full shift, for example, has repelling periodic points of all orders (Module 13, Exercise 9c), and we have seen that the map f defined on $\mathbb{R}$ in Module 4 also has repelling periodic points of all orders. How closely are $(\mathbb{R}, f)$ and $(\{0,1\}^{\mathbb{N}}, \sigma)$ related? Is it possible that they are conjugate?

In Module 8 we saw that a conjugacy between two systems is a homeomorphism between the phase spaces that commutes with the maps. Thus a conjugacy must be a continuous function. As you might have observed in the exploration, the intuition and skills that you have developed in calculus for determining whether or not a function is continuous do not translate well into the symbolic setting. Thus we will need a more rigorous understanding of continuity before we will be able to construct conjugacies involving symbolic dynamical systems.

The formal definition of continuity involves sequences of points. Recall from Exercise 5 in Module 13 that describing a sequence of points in a shift space is notationally tricky because the points are themselves sequences. In order to avoid confusion, we use the notation $\{x^{(n)}\}_{n=1}^{\infty}$ for a sequence of points in a shift space.

This notation will help us distinguish a point $x^{(n)}$ in the sequence of points from the symbol x_m in the mth position of the point x. So, for example, $x_m^{(n)}$ is the symbol in the mth position of the nth point in the sequence.

We state the definition of continuity using this notation:

Definition 14.1. *Let X and Y be metric spaces. A function $f : X \to Y$ is* **continuous** *if for all $a \in X$ and any sequence $\{x^{(n)}\}_{n=1}^{\infty}$ of points in X that converge to a, we have*

$$\lim_{n\to\infty} f(x^{(n)}) = f(a).$$

In preparation for questions about conjugacy, we will explore the continuity of functions that have a shift space for their domain or range (or both). We will do this by looking at three examples, the first of which is a function whose domain and range are both $\{0,1\}^{\mathbb{Z}}$.

Example 1: Consider the function $\phi : \{0,1\}^{\mathbb{Z}} \to \{0,1\}^{\mathbb{Z}}$ where $\phi(x) = y$ is defined by the rule

$$y_i = (x_i + x_{i+1}) (\text{mod } 2).$$

That is, the symbol in the ith position of y is determined by the symbols in the ith and $(i+1)$st positions in x. For example,

$$\text{if} \quad x = \ldots 000.1101\ldots \quad \text{then} \quad \phi(x) = \ldots 001.011\ldots.$$

We can see that the function ϕ is not one-to-one since

$$\phi(\ldots 000.000\ldots) = \phi(\ldots 111.111\ldots) = \ldots 000.000\ldots.$$

It is, however, onto (Exercise 2). In order to show that ϕ is continuous we must show that given $a \in \{0,1\}^{\mathbb{Z}}$ and a sequence $x^{(n)}$ of points in $\{0,1\}^{\mathbb{Z}}$ converging to a. the sequence of points $\phi(x^{(n)})$ converges to $\phi(a)$. In $\{0,1\}^{\mathbb{Z}}$, the fact that $x^{(n)}$ converges to a means that given k, we can choose N large enough so that $x_{[-k,k]}^{(n)} = a_{[-k,k]}$ whenever $n \geq N$. We must show that the same holds true for $\phi(x^{(n)})$ and $\phi(a)$.

So, let k' be given. Let $k = k' + 1$ and choose N large enough that $x^{(n)}_{[-k,k]} = a_{[-k,k]}$ whenever $n \geq N$. By the definition of ϕ, the entry in the ith position in the sequence $\phi(x^{(n)})$ is determined by the entries in the ith and $(i+1)$st positions of $x^{(n)}$. Thus, if $x^{(n)}_{[-k,k]} = a_{[-k,k]}$ then $\phi(x^{(n)})_{[-k,k-1]} = \phi(a)_{[-k,k-1]}$. So, whenever $n \geq N$, $\phi(x^{(n)})_{[-k',k']} = \phi(a)_{[-k',k']}$. This tells us that

$$\lim_{n\to\infty} \phi(x^{(n)}) = \phi(a)$$

as desired, and ϕ is continuous.

Finding continuous functions between a shift space and $\mathbb{R}$ (or a subset of $\mathbb{R}$) is more complicated. We will illustrate the complications with two examples. The use of the binary expansion of numbers in $[0, 1]$ will be helpful, so we start by reviewing this idea.

Definition 14.2. *Let $z \in [0, 1]$. A sequence $.z_0z_1\ldots$ is called a* **binary expansion** *of z if $z_i = 0$ or 1 for all $i = 0, 1, \ldots$ and*

$$z = \sum_{i=0}^{\infty} z_i \cdot \frac{1}{2^{i+1}}.$$

Every $z \in [0, 1]$ has a binary expansion. For example, as you will show in Exercise 4, $1/3$ has the binary expansion $.010101\ldots$ since

$$\frac{1}{3} = 0 \cdot \frac{1}{2} + 1 \cdot \frac{1}{2^2} + 0 \cdot \frac{1}{2^3} + 1 \cdot \frac{1}{2^4} + \cdots.$$

We call this a **nonterminating binary expansion** since the pattern observed on the terms shown above continues and there are infinitely many nonzero terms in the sum. The binary expansion of $1/3$ is unique (Exercise 5), but some numbers have exactly two binary representations. For example (Exercise 4), $1/2$ has as its binary representation either $.10000\ldots$ since

$$\frac{1}{2} = 1 \cdot \frac{1}{2} + 0 \cdot \frac{1}{2^2} + 0 \cdot \frac{1}{2^3} + 0 \cdot \frac{1}{2^4} + \cdots$$

or $.01111\ldots$ since

$$\frac{1}{2} = 0 \cdot \frac{1}{2} + 1 \cdot \frac{1}{2^2} + 1 \cdot \frac{1}{2^3} + 1 \cdot \frac{1}{2^4} + \cdots.$$

The first representation of $1/2$ is called the **terminating binary expansion** because there are only finitely many nonzero terms. The second representation is its nonterminating expansion. In Exercise 6 you will show that if $z \in [0, 1]$ has two binary expansions then it has one terminating and one nonterminating expansion. The word terminating describes the fact that there are only finitely many nonzero terms in the sum and thus we can write it as a finite sum; we can still think of every binary expansion as giving us an infinite sequence of 0s and 1s, with a terminating expansion having the property that $z_i = 0$ for all i after a certain value.

The first of our next two examples will consist of a function that maps a point x in $\{0, 1\}^{\mathbb{N}}$ to the point $z \in [0, 1]$ with $.x_0x_1\ldots$ as its binary expansion. The function in the second example will be a type of inverse for the first function: that is, it will map $z \in [0, 1]$ to the one-sided sequence of 0s and 1s in its binary expansion. As we will see, the continuity properties for these two examples will be very different.

Example 2: Consider the function $f : \{0, 1\}^{\mathbb{N}} \to \mathbb{R}$ defined as follows: for any

$$x = .x_0x_1x_2x_3 \ldots \in \{0, 1\}^{\mathbb{N}}$$

let

$$f(x) = x_0 \cdot \frac{1}{2} + x_1 \cdot \frac{1}{2^2} + x_2 \cdot \frac{1}{2^3} + \ldots .$$

It is not difficult to see that f is neither one-to-one nor onto (Exercise 7).

In order to explore the continuity of f, let [0] and [1] denote the set of points $x \in \{0, 1\}^{\mathbb{N}}$ with $x_0 = 0$ and $x_0 = 1$ respectively. That is, [0] is the set of all one-sided sequences in $\{0, 1\}^{\mathbb{N}}$ with a 0 in the zeroth position, and [1] is the set of all one-sided sequences in $\{0, 1\}^{\mathbb{N}}$ with a 1 in the zeroth position. If $x \in [0]$, then

$$f(x) = 0 \cdot \frac{1}{2} + x_1 \cdot \frac{1}{4} + x_2 \cdot \frac{1}{8} + \ldots \leq \frac{1}{2},$$

and if $x \in [1]$, then

$$f(x) = 1 \cdot \frac{1}{2} + x_1 \cdot \frac{1}{4} + x_2 \cdot \frac{1}{8} + \ldots \geq \frac{1}{2}.$$

Thus points in [0] must get mapped by f to the interval $[0, 1/2]$ and points in [1] must get mapped by f to the interval $[1/2, 1]$, as indicated in Figure 14.1.

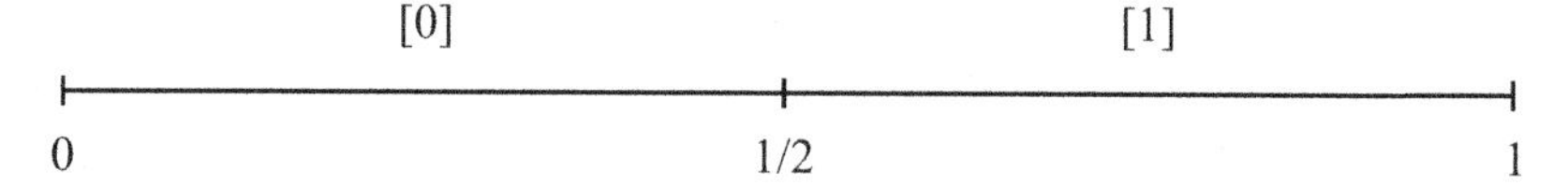

Figure 14.1. The image under f of the points in the sets [0] and [1].

Next we define $[00] = \{x \mid x_0x_1 = 00\}$. If $x \in [00]$, then

$$f(x) = 0 \cdot \frac{1}{2} + 0 \cdot \frac{1}{4} + x_2 \cdot \frac{1}{8} + \ldots \leq \frac{1}{4}.$$

So, f maps the set [00] to the interval $[0, 1/4]$. If we define the subsets [01], [10], and [11] of $\{0, 1\}^{\mathbb{N}}$ analogously, we see that they map to the intervals shown in Figure 14.2.

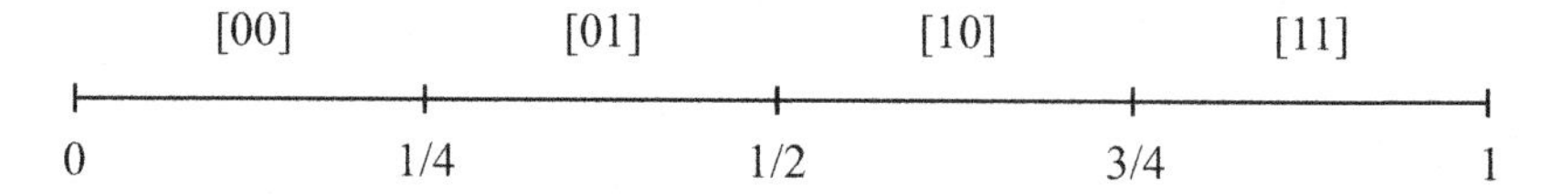

Figure 14.2. The image under f of the points in the sets [00], [01], [10], and [11].

We can continue with this reasoning to show that if two points $x, y \in \{0, 1\}^{\mathbb{N}}$ agree for a block of size k starting at the origin, then their images must lie in an interval of size $1/2^k$. In Exercise 7 you will prove this fact more formally in order to show that f is a continuous function.

Example 3: Consider the function $g : [0, 1] \to \{0, 1\}^{\mathbb{N}}$ defined as follows: let $z \in [0, 1]$ have binary expansion

$$z = \sum_{i=0}^{\infty} z_i \cdot \frac{1}{2^{i+1}}.$$

Then define

$$g(z) = .z_0 z_1 z_2 z_3 \dots.$$

If z has two binary expansions, then we will agree to use the terminating expansion of z when defining g. For example, $g(1/2) = .1000\dots$.

The function g is one-to-one, but it is not onto (Exercise 8). It is also not continuous, and the discontinuity is caused exactly by those numbers $a \in [0, 1]$ with two binary expansions. For example, let $a = 1/2$. As discussed earlier, in addition to its terminating binary expansion, $a = 1/2$ has the nonterminating binary expansion

$$\frac{1}{2} = 0 \cdot \frac{1}{2} + 1 \cdot \frac{1}{2^2} + 1 \cdot \frac{1}{2^3} + 1 \cdot \frac{1}{2^4} + \dots;$$

by taking $z^{(1)} = \frac{1}{2^2}$, $z^{(2)} = \frac{1}{2^2} + \frac{1}{2^3}$, $z^{(3)} = \frac{1}{2^2} + \frac{1}{2^3} + \frac{1}{2^4}$, and so on, we will be able to find a sequence of numbers in $[0, 1]$ whose binary expansions converge to the nonterminating expansion of $a = 1/2$. This is illustrated in Figure 14.3:

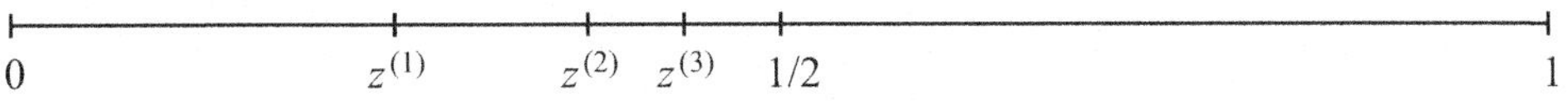

$0 \qquad z^{(1)} \qquad z^{(2)} \quad z^{(3)} \quad 1/2 \qquad 1$

Figure 14.3. The first few terms of the sequence $z^{(n)}$ positioned on the number line.

Formally, let

$$z^{(n)} = \sum_{i=1}^{n} \frac{1}{2^{i+1}}$$

and notice that

$$\begin{aligned}
\left|\frac{1}{2} - z^{(n)}\right| &= \left|\frac{1}{2} - \left(\frac{1}{4} + \frac{1}{8} + \dots + \frac{1}{2^{n+1}}\right)\right| \\
&= \left|\left(\frac{1}{4} + \frac{1}{8} + \dots + \frac{1}{2^{n+1}} + \frac{1}{2^{n+2}} + \frac{1}{2^{n+3}} + \dots\right)\right. \\
&\qquad \left. - \left(\frac{1}{4} + \frac{1}{8} + \dots + \frac{1}{2^{n+1}}\right)\right| \\
&= \frac{1}{2^{n+2}} + \frac{1}{2^{n+3}} + \dots .
\end{aligned}$$

In Exercise 4c you will show this equals $1/2^{n+1}$, and thus $\lim_{n\to\infty} z^{(n)} = 1/2$. On the other hand,

$$g(z^{(n)}) = .01^n\,0\,0\dots$$

where 1^n denotes n consecutive occurrences of the symbol 1, and

$$\lim_{n\to\infty} g(z^{(n)}) = .011111\dots.$$

This does not equal $g(a) = g(1/2) = .1000\dots$.

Since we have found a sequence $\{z^{(n)}\}_{n=1}^{\infty}$ in $[0, 1]$ that converges to a, but for which $\{g(z^{(n)})\}_{n=1}^{\infty}$ does not converge to $g(a)$, we have shown that g is not continuous at $a = 1/2$.

We did have a choice initially of which expansion, the terminating or the nonterminating, to use in defining the function g. In Exercise 9 you will show that using the nonterminating expansion also results in a discontinuous function.

The following theorem tells us, perhaps surprisingly, that the failure of continuity for the function g is no accident.

Theorem 14.3. *Let S be a subset of $\{0,1\}^{\mathbb{Z}}$ or $\{0,1\}^{\mathbb{N}}$. If $g : [0,1] \to S$ is continuous, then g is a constant function.*

The proof of this theorem would take us too far off topic and so we omit it. However, for those students who have taken analysis and topology, the theorem follows immediately from two facts: the continuous image of a connected set must be connected and any subset of $\{0,1\}^{\mathbb{Z}}$ or $\{0,1\}^{\mathbb{N}}$ is totally disconnected.

Theorem 14.3 tells us that there are no homeomorphisms from $[0,1]$ to a shift space, and thus there can be no conjugacy between any dynamical system whose phase space is $[0,1]$ and any symbolic dynamical system. However, there are homeomorphisms between subsets of $[0,1]$ and shift spaces. In Exercise 11 you will prove the following theorem.

Theorem 14.4. *Let $\Gamma \subset [0,1]$ be the Cantor set. Then Γ is homeomorphic to $\{0,1\}^{\mathbb{N}}$.*

The object of study in Module 4 was a dynamical system (Γ, f) with the Cantor set Γ as its phase space, so the existence of a homeomorphism between Γ and $\{0,1\}^{\mathbb{N}}$ raises the question as to whether (Γ, f) and $(\{0,1\}^{\mathbb{N}}, \sigma)$ are conjugate dynamical systems. It is tempting to think that conjugacy follows immediately from Theorem 14.4, but the definition of a conjugacy requires finding a homeomorphism that also commutes with f and σ. There are in fact many homeomorphisms between Γ and $\{0,1\}^{\mathbb{N}}$, and the most natural ones will not be conjugacies. However, as you will see in the project, it is possible to find a homeomorphism between (Γ, f) and a symbolic dynamical system that is also a conjugacy.

Exercises

1. Are the functions f and g defined in the exploration continuous? Prove your claims.

2. Show that if $\phi : \{0,1\}^{\mathbb{Z}} \to \{0,1\}^{\mathbb{Z}}$ is the function defined in Example 1, then every sequence $y \in \{0,1\}^{\mathbb{Z}}$ has exactly two preimages. That is, for any $y \in \{0,1\}^{\mathbb{Z}}$, show that there exist exactly two distinct sequences $x, z \in \{0,1\}^{\mathbb{Z}}$ with $\phi(x) = \phi(z) = y$.

3. The function ϕ from Example 1 could also be used as a function from $\{0,1\}^{\mathbb{N}}$ to $\{0,1\}^{\mathbb{N}}$. Discuss whether ϕ is continuous in this setting.

4. Recall that a geometric series has the form

$$a + ar + ar^2 + \cdots + ar^n + \ldots$$

and that if $|r| < 1$ then it converges to $a/(1-r)$. Use this to show

(a) $0 \cdot \frac{1}{2} + 1 \cdot \frac{1}{2^2} + 0 \cdot \frac{1}{2^3} + 1 \cdot \frac{1}{2^4} + \ldots = \frac{1}{3}$

(b) $0 \cdot \frac{1}{2} + 1 \cdot \frac{1}{2^2} + 1 \cdot \frac{1}{2^3} + 1 \cdot \frac{1}{2^4} + \ldots = \frac{1}{2}$

(c) $\frac{1}{2^{n+2}} + \frac{1}{2^{n+3}} + \ldots = \frac{1}{2^{n+1}}$.

5. Show that $1/3$ has a unique binary expansion.

6. Show that if $z \in [0, 1]$ has two binary expansions then one of them must be terminating and the other must be nonterminating.

7. This question concerns the function $f : \{0, 1\}^{\mathbb{N}} \to \mathbb{R}$ defined in Example 2.

   (a) Prove that f is neither one-to-one nor onto.

   (b) Prove that f is a continuous function.

8. Show that the function $g : [0, 1] \to \{0, 1\}^{\mathbb{N}}$ defined in Example 3 is one-to-one but not onto.

9. Modify the definition in Example 3 to use the nonterminating expansion in the case where z has two binary expansions. Show that this newly defined function is discontinuous.

10. (a) Show that the shift map $\sigma : \{0, 1\}^{\mathbb{Z}} \to \{0, 1\}^{\mathbb{Z}}$ is a continuous map.

    (b) Find an example of a function $\alpha : \{0, 1\}^{\mathbb{Z}} \to \{0, 1\}^{\mathbb{Z}}$ that is not continuous. Is your example one-to-one? Onto? Prove your assertions.

    (c) Do your answers change if we use $\{0, 1\}^{\mathbb{N}}$ instead of $\{0, 1\}^{\mathbb{Z}}$?

11. Let Γ be the Cantor set. Show that Γ is homeomorphic to $\{0, 1\}^{\mathbb{N}}$.

12. Let X_1 be the shift space with alphabet $\mathcal{A}_1 = \{0, 1, 2\}$ and forbidden blocks $\mathcal{F}_1 = \{01, 20, 12\}$ and let X_2 be the shift space with alphabet $\mathcal{A}_2 = \{a, b, c\}$ and forbidden blocks $\mathcal{F}_2 = \{ab, ca, bc\}$. Show that (X_1, σ) and (X_2, σ) are conjugate.

13. Is the full shift on two symbols conjugate to the full shift on three symbols?

14. Show that $(\{0, 1\}^{\mathbb{Z}}, \sigma)$ is not conjugate to $(\{0, 1\}^{\mathbb{N}}, \sigma)$.

15. Let X be the golden mean shift as defined in Module 13, Exercise 1. Let $Y \subset \{0, 1, 2\}^{\mathbb{Z}}$ be given by forbidden blocks $\mathcal{F} = \{02, 10, 11, 22\}$. Define a map $\phi : X \to Y$ by

$$\phi(x)_i = \begin{cases} 0 & \text{if} \quad x_i x_{i+1} = 00 \\ 1 & \text{if} \quad x_i x_{i+1} = 01 \\ 2 & \text{if} \quad x_i x_{i+1} = 10. \end{cases}$$

Show that (X, σ) is conjugate to (Y, σ).

Project

Let $(\mathbb{R}, f)$ be the dynamical system given by

$$f(x) = \begin{cases} 3x & \text{if } x \leq \frac{1}{2} \\ 3 - 3x & \text{if } x > \frac{1}{2}. \end{cases}$$

Let Γ be the set of initial values with bounded orbits as defined in Module 4. Associate to each $x \in \Gamma$ a one-sided sequence of the form $\phi(x) = .x_0x_1x_2\ldots$ by setting, for each $i \geq 0$,

$$x_i = \begin{cases} 0 & \text{if } f^i(x) \in [0, \frac{1}{3}] \\ 1 & \text{if } f^i(x) \in [\frac{2}{3}, 1] \end{cases}$$

where $f^0(x) = x$. The sequence $\phi(x)$ associated with $x \in \Gamma$ is called its **itinerary**. Describe the set X of one-sided sequences that arise as itineraries of points in Γ. Show that ϕ a conjugacy between (Γ, f) and (X, σ).

Further Reading

There are many excellent books accessible to undergraduates where the interested reader can learn more about dynamical systems. We do not attempt to provide a complete list of references, but we make a few suggestions for further reading in the topics covered in this text. The books by Bob Devaney [D1], [D2]; by Alligood, Sauer, and Yorke [ASY]; and by Denny Gulick [G] all provide a thorough treatment of many of the topics. A staple in the study of symbolic dynamical systems is the book by Lind and Marcus [LM]. References specific to particular modules are listed and described below.

Module 5

There are multiple definitions of chaos in the literature. The definition of chaos used in the module can be found in Devaney [D1]. The article [MDS] describes a variety of other definitions and analyzes their key components. It provides many examples and a chart (Table 3.1 in [MDS]) for easy comparison.

We have restricted our attention to functions defined on intervals for much of this text, and in this situation the Devaney definition of chaos can be simplified. In [VB], a short, simple proof establishes the fact that if I is an interval and $f : I \to I$ is continuous and transitive, then f is chaotic. We use the original Devaney definition in the module both because it better describes the important features of a chaotic system and because it can be generalized to other situations.

Module 6

The original source (published in Russian) for Sharkovskii's theorem is [S]. A special case of Sharkovskii's theorem, published by Li and Yorke without knowledge of Sharkovskii's work, can be found in [LY]. They prove that if I is an interval and continuous function $f : I \to I$ has a 3-cycle, then f has a period k point for all $k \in \mathbb{N}$. The proof of this result is fairly short and requires only basic facts from analysis. A thorough and readable proof of the full Sharkovskii theorem can be found in [BH]. Section 1.2 of [BH] also includes a nice history of Sharkovskii's theorem and its various proofs.

Module 10

Iterated function systems are the subject of many engaging books, articles, and websites, and an interested reader can have great fun exploring. We mention three starting points.

A treatment of iterated function systems by viewing them as functions taking sets to sets, consistent with the discussion preceding Theorem 10.4 in Module 10, is found in Section 7 of Chapter 3 of [Ba]. (Chapter 2 is a prerequisite.)

The textbook [F] takes a geometric view of fractals. It is a graduate text, but an advanced undergraduate may find it useful and any reader can enjoy the discussion of fractals found in its introduction. A more detailed discussion of fractals can be found in Chapters 2 and 3. Iterated function systems are covered in Chapters 9 and 13. Many of the results in Module 10 are contained in Theorem 9.1.

Another reference appropriate for the advanced undergraduate is the research article [H]. Section 3 of this article generalizes the results found in the module.

Module 11

Chapter 3 of [Ba] introduces Möbius transformations, and Chapter 4 gives a treatment of their basic properties. Another useful textbook for this topic is [Be]. Although [Be] is written for graduate students, Chapter 1 discusses many examples, including the Möbius transformations, at a level accessible to advanced undergraduates.

Module 12

There are many websites that yield beautiful images of the Julia and Mandlebrot sets, and many give nice descriptions of the mathematics behind the images. One can also find a discussion of the Julia and Mandelbrot sets in Chapter 14 of [F], and a thorough description of them is given in Chapters 7 and 8 of [Ba].

Module 14

The book by Lind and Marcus [LM] contains other uses and examples of conjugacies. Additional accessible examples can be found in the papers of Bernhardt and Yuster [BY] and Johnson and Madden [JM].

Bibliography

[ASY] Alligood, K., Sauer, T., and Yorke, J. *Chaos: an Introduction to Dynamical Systems*, Springer-Verlag, 1996.

[Ba] Barnsley, M. *Fractals Everywhere*, Academic Press, 1993.

[Be] Beardon, A. *Iteration of Rational Functions*, Springer-Verlag, 1991.

[BY] Bernhardt, C. and Yuster, T. "Periodic Points of the Difference Operator," *The College Mathematics Journal*, 28 **1** (1997) 20–26.

[BH] Burns, K. and Hasselblatt, B. "The Sharkovsky Theorem: A Natural Direct Proof," *American Mathematical Monthly*, 118 **3** (2011) 229–244.

[D1] Devaney, R. *A First Course in Chaotic Dynamical Systems*, Westview Press, 1992.

[D2] Devaney, R. *An Introduction to Chaotic Dynamical Systems*, Westview Press, 2003.

[E] Eloranta, K. "A Note on Certain Rigid Subshifts," *London Mathematical Society Lecture Notes Series*, 228 (1996) 307–317.

[EH] Eisele, P. and Hadler, K.P. "Game of Cards, Dynamical Systems, and a Characterization of the Floor and Ceiling Functions," *American Mathematical Monthly*, 97 (1990) 466–477.

[F] Falconer, K. *Fractal Geometry: Mathematical Foundations and Applications*, John Wiley and Sons, 2003.

[FJS] Frame, M., Johnson, B. and Sauerberg, J. "Fixed Points and Fermat: A Dynamical Systems Approach to Number Theory," *American Mathematical Monthly*, 107 (2000) 422–428.

[G] Gulick, D. *Encounters with Chaos*, McGraw-Hill, 1992.

[H] Hutchinson, J. "Fractals and Self-Similarity," *Indiana University Math. Journal*, 30 **5** (1981) 713–747.

[JM] Johnson, A. and Madden, K. "Renewal Systems, Sharp-Eyed Snakes, and Shifts of Finite Type," *American Mathematical Monthly*, 109 **3** (2002) 258–272.

[LY] Li, T. and Yorke, J. "Period Three Implies Chaos," *American Mathematical Monthly*, 82 **10** (1975) 985–992.

[LM] Lind, D. and Marcus, B. *An Introduction to Symbolic Dynamics and Chaos*, Cambridge University Press, 1995.

[MDS] Martelli, M., Dang, M. and Seph, T. "Defining Chaos," *Mathematics Magazine*, 71 **2** (1998) 112–122.

[S] Sharkovskii, O. "Co-existence of Cycles of a Continuous Map of the Line onto Itself," *Ukrainskii Matematicheskii Zhurnal*, 16 **1** (1964) 61–71.

[VB] Vellekoop, M. and Berglund, R. "On Intervals, Transitivity Equals Chaos," *American Mathematical Monthly*, 101 **4** (1994) 353–355.

[Wa] Walsh, J. "The Dynamics of Newton's Method for Cubic Polynomials," *College Mathematics Journal*, 26 **4** (1995) 22–28.

[Wi] Wiles, A. "The Proof," (1997) www.pbs.org/wgbh/nova/proof/

Index